*Applications of HPLC
in biochemistry*

D1536993

LABORATORY TECHNIQUES IN BIOCHEMISTRY AND MOLECULAR BIOLOGY

Volume 17

Edited by

R.H. BURDON—*Department of Bioscience and Biotechnology, University of Strathclyde, Glasgow*
P.H. van KNIPPENBERG—*Department of Biochemistry, University of Leiden, Leiden*

Advisory board

P. BORST—*University of Amsterdam*
D.C. BURKE—*Allelix Inc., Ontario*
P.B. GARLAND—*University of Dundee*
M. KATES—*University of Ottawa*
W. SZYBALSKI—*University of Wisconsin*
H.G. WITTMAN—*Max-Planck Institut für Molekuläre Genetik, Berlin*

ELSEVIER
AMSTERDAM · NEW YORK · OXFORD

APPLICATIONS OF HPLC
IN BIOCHEMISTRY

A. Fallon
British Bio-technology Limited
Brook House, Watlington·Road
Oxford OX4 5LY, U.K.

R.F.G. Booth
The Wellcome Research Laboratories
Langley Court, Beckenham
Kent BR3 3BS, U.K.

and

L.D. Bell
Monsanto Company
700 Chesterfield Village Parkway
St. Louis, MO 63198, U.S.A.

ELSEVIER
AMSTERDAM · NEW YORK · OXFORD

ISBN 0-444-80863-9 (pocket edition)   1st edition   1987
ISBN 0-444-80862-0 (library edition)   2nd printing   1988
ISBN 0-7204-4200-1 (series)   3rd printing   1990

Published by:
ELSEVIER SCIENCE PUBLISHERS B.V. (BIOMEDICAL DIVISION)
P.O. BOX 211
1000 AE AMSTERDAM
THE NETHERLANDS

Sole distributors for the U.S.A. and Canada:
ELSEVIER SCIENCE PUBLISHING COMPANY, INC.
52 VANDERBILT AVENUE
NEW YORK, NY 10017
U.S.A.

Library of Congress Card No. 85-647011

Printed in the Netherlands

"There is more to life than increasing its speed"

Mahatma Gandhi (1869–1948)

Acknowledgements

The authors would like to acknowledge the people who have contributed to this book. In particular Mrs. Philippa M. Bell and Miss Patricia Rudge for illustrations and their support, Mrs. R.A. Fallon for being there, and Mr. David Flack for illustrations. In addition the authors gratefully acknowledge the contribution of Mrs. Joan Budgen, Maureen Bevan and Mrs. Helena Grant for excellent typing services.

One final recognition is the contribution of Phillip K. Branston of Anachem Ltd. for his enthusiastic sales services without which this book may have never been written.

Contents

Chapter 6. High performance normal phase chromatography 65

Chapter 7. High performance reversed phase chromatography ... 73

Chapter 11. Applications of HPLC

Chapter 11.1. Nucleosides and nucleotides *144*

CHAPTER 1

The origins and development of liquid chromatography

1.1. Origins

The history of liquid chromatography and the influence of its pioneers have been the subject of several learned reviews (Sakodynskii, 1970; Kirchner, 1973; Ettre and Horvath, 1975). Indeed, the manner in which scientific attitudes and the turbulent history of Europe in the early decades of this century influenced the establishment of chromatography as a science would in itself be worthy of a textbook. In a book of this nature a detailed historical narrative would not be relevant but it is instructive to put the current state of the art of liquid chromatography within a historical perspective to provide the reader with a realisation of the explosive development of chromatographic science within the past fifteen years.

The earliest example of the use of chromatography to elicit a separation is credited to the highly gifted Russian botanist Michael Tswett who, in a period between 1903 and 1906, used adsorption chromatography on a calcium carbonate column to separate various plant pigments from leaf extracts. The further exploration of chromatographic science remained almost dormant until, in 1931, the technique was rediscovered (Kuhn et al., 1931). The cause of the twenty-year period of stagnation has been extensively reviewed by Ettre and Horvath (1975), who refer to the dominance of European organic chemistry by specific German universities where the research efforts were directed more towards the development of industrial chemistry. At this time the techniques used for the preparation and purification of industrial chemicals were distillation and crystallisation and the potential of chromatography remained unrealised, resulting in a general neglect of Tswett's work. In 1931 the first of a series of articles originating from the laboratory of Richard Kuhn was published (Kuhn et al., 1931) which described the use of adsorption chromatography to resolve from the 'lutein' of egg yolk two different carotenes. Within a short period many other laboratories had adopted the technique. Rapid expansion followed and in 1941 partition chromatography was described (Martin and Synge, 1941) and subsequently paper chromatography (Consden et al., 1944) which revolutionised approaches to biochemical analysis.

The basic principles underlying the technique of liquid chromatography were thereby established and within a short time span advances in both the theoretical and practical aspects of the science resulted in its rapid popularisation as the method of choice for the separation and quantitation of the components of unknown mixtures.

1.2. Recent advances in liquid chromatography

In this book we shall confine ourselves to the theory and practice of HPLC, which has been variously described as high performance liquid chromatography, high pressure liquid chromatography and, more recently, as highly priced liquid chromatography; the reader will have to decide for himself the truth of the latter! It is in the field of HPLC that the major technical advances have recently occurred resulting in the widespread popularisation of liquid chromatography. Several practical problems hindered the general applicability of liquid chromatography in both analytical and preparative modes and it is these we shall now consider together with the developments which provided solutions.

The analytical columns used in liquid chromatography were until recently of somewhat variable quality and in addition few column packing materials were available, thereby restricting the choice of both the solid phase (i.e. column packing material) and the mobile phase (i.e. the eluent). Also, to allow a good resolution of chemically similar compounds, the columns needed to be rather long, thus resulting in extended elution times with a concomitant consumption of large quantities of mobile phase. Columns with very high resolving power are now commercially available and with careful usage possess lifetimes of at least 12 months. The range of available packing materials is vast and will facilitate most separations. Furthermore, the techniques now available allow reproducible columns to be self-packed at a fraction of the cost of commercially supplied columns. In general the chromatography columns now available, whether self-packed or commercially supplied, contain the equivalent resolving power within a few column centimetres to that found previously in conventional liquid chromatography within metres. This has resulted in much reduced elution times such that most separations are now achieved within thirty minutes, saving both time and solvent. The developments in the quality and variety of column packing materials and the advances in column packing technology has contributed perhaps more than any other single factor to the popularisation of HPLC.

Originally, liquid chromatography relied upon gravity for the delivery of the mobile phase to the column, but this was superseded by the use of various types of pumps (or, to use the more common technical jargon, 'solvent delivery modules') to deliver the mobile phase. Pump technology has undergone several recent advances to cope with the increasingly sophisticated demands of modern HPLC. Pumps are now frequently under microprocessor control and are able to accurately deliver solvent over a broad range of flow rates (0.1 μl/min–20 ml/min). Gradient elution has allowed HPLC to be used for chromatographic separations where the use of only one mobile phase (isocratic separation) would not have achieved adequate resolution. The intervention of microprocessor technology has allowed the generation of extremely sophisticated gradients to reproducibly facilitate very complex separations. Recent advances in the control of solvent delivery at very low flow rates has also allowed the development of microbore HPLC systems which will be discussed later in the book. Additionally, the improvements in pump technology have resulted in essentially non-pulsatile flow, a development of significance in determining the lower limits of detector sensitivity.

The advances in column and pump technology and the introduction of microprocessor controls have allowed HPLC to become accessible as an analytical tool to individuals not specifically devoted to the technique and it is this expansion in availability which has resulted in an upsurge in the popularity of HPLC such that it is now commonly the method of choice in chromatographic analysis.

1.3. The advantages of HPLC over other chromatographic techniques

An outline of some of the reasons for the recent popularisation of HPLC has been given, but it should be realised that while the advent of HPLC has resulted in completely new applications where no alternatives were previously available, many of the uses to which HPLC is now put have succeeded other chromatographic techniques. In this section HPLC will be compared with thin layer chromatography (TLC), gas liquid chromatography (GLC) and paper chromatography.

The single most significant advantage of HPLC resides in the variety of different chromatographic techniques readily available, including partition chromatography, adsorption chromatography, ion-exchange chromatography, size exclusion chromatography, reversed phase and normal phase bonded chromatography and affinity chromatography. The majority of the column packings necessary for the different types of chromatography have only just become accessible to HPLC as a consequence of the developments in column technology referred to earlier. The vast array of column materials available within each of the different chromatographic modes may be contrasted with the comparative paucity of solid phases available in TLC, GLC or paper chromatography where, even if the appropriate chromatographic mode exists, the separations can usually only be optimised by alterations of the mobile phase.

In HPLC the mobile phase is also easily manipulated; indeed, the use of mobile phase gradient systems provides HPLC with one of its greatest advantages over other chromatographic methods. In GLC no equivalent technique exists, while in TLC and paper chromatography a change in the mobile phase usually requires the removal of the solid phase from its chromatographic tank. However, none of these techniques match the sophistication or ease with which mobile phase gradients may be generated in HPLC.

The introduction of reversed phase chromatography and its affiliated technique of ion-pairing has provided HPLC with one of its most useful methodologies, allowing the resolution of both very similar and completely different compounds on the same chromatographic system. In most other forms of chromatography either the mobile phase or the stationary phase needs to be altered to resolve wholly different chemical species. Recently reversed phase TLC has been developed which provides some of the advantages found in reversed phase HPLC.

The main rival of HPLC for chromatographic separations is GLC. However, for a sample to be suitable for GLC analysis it must be either thermally volatile or subjected to a derivatisation procedure to enhance its volatility. Derivatisation procedures introduce several inconveniences. Firstly, sample derivatisation is a lengthy, time-consuming procedure; secondly, a derivatisation procedure may not be possible or the process of derivatisation may degrade the sample; and finally, derivatisation introduces another potential source of quantitative error. Derivatisation is also necessary in GLC where electron capture detection systems are utilised since the sample to be analysed needs to be coupled to a specific electron-absorbing reagent. In contrast, more than 70% of HPLC separations are performed using UV detectors in which derivatisation is not necessary. Of the remaining 30% of analyses, approximately half rely upon fluorescent detection which may or may not require derivatisation. In general the sole requirements of a sample for HPLC is that it is soluble in a solvent which will not damage the column nor interfere with the detector response.

Since HPLC is non-destructive the resolved components of a mixture may be collected, thus allowing HPLC to be used as a preparative technique. Indeed, the use of long, wide diameter columns makes the purification of gram quantities of material a relatively facile process. Moreover, the absence of sample modification throughout preparation and detection allows more than one type of detector to be connected in series; for example, the use of a UV detector connected to an electrochemical detector considerably enhances the potential for characterisation of unknown components eluting from a column.

The popularity and relative ease of use of UV detection systems in HPLC should not obscure the multitude of detectors which are now commercially available, including both fixed and variable wavelength UV detectors, fluorimeters, refractive index detectors, amperometric detectors, radioactive flow detectors, coulombometric detectors and several other detection systems which are still at a comparatively early stage of development. The details of these different systems will be discussed later but inherent advantages over the detection techniques used in paper chromatography and TLC are that they provide both continuous monitoring of a chromatogram and may also be directly interfaced with computer hardware to provide data processing. In both paper chromatography and TLC accurate quantitation is a relatively time-consuming process.

The time required for most HPLC separations is less than thirty minutes and with some high speed HPLC separations which will be described later these times may be further reduced to less than two minutes. While GLC can often match this performance, neither TLC nor paper chromatography can provide resolution in a comparable time. Apart from the obvious advantage of time saving, relatively labile samples may be rapidly analysed by HPLC before any deterioration has occurred. The problem of sample lability may also be a problem in GLC where thermally labile compounds may be decomposed by the relatively high temperature used in GLC analyses. In contrast, most HPLC separations are carried out at room temperature.

Finally, with the high cost of capital outlay for sophisticated chromatographic equipment and the associated expense of a trained operator, any innovation which provides increased cost-effectiveness is an essential development. Recently, substantial advances have been made in the automation of HPLC to facilitate the reliable overnight operation of HPLC systems. Similar developments have been made in the automation of GLC but the technical advances required for similar operations to occur in either TLC or paper chromatography analysis would seem to preclude complete automation of these systems in the forseeable future.

1.4. Outline of the book

It is the intention of this text to provide a practical guide to HPLC for both the novice and the chromatographer already using the technique who is seeking new methodologies or applications. A chapter is included on the theory of chromatography which describes the fundamental principles underlying the practice of HPLC with the emphasis heavily oriented toward the use of chromatography theory to elicit good chromatographic separations. Subsequent chapters describe the main chromatographic techniques used in present day HPLC and once again the emphasis is on the practical application of these various techniques and the way in which variable parameters, such as ionic strength, pH etc., will affect a chromatographic separation.

One chapter is devoted to a description of the basic instrumentation used in HPLC, together with a general review of the specific applications of different equipment in various types of chromatography. Also a 'state of the art' resumé together with some general recommendations are made regarding commercially available hardware. A chapter on various practical aspects of HPLC gives advice on such matters as sample handling, establishing a separation, choice of separation mode, choice of solvents etc. Finally, a chapter of some of the different applications of HPLC has been included which of necessity is restricted to certain specific groups of compounds.

Examples of actual chromatograms together with the details of the stationary and mobile phases are presented with the intention that, even if the reader cannot find the particular separation which is required, the information provided will allow appropriate conditions to be devised.

CHAPTER 2

The theory of HPLC

2.1. Introduction

Liquid chromatography is a separation method in which a mixture of components is resolved into its constituent parts by passage through a chromatographic column. It is carried out by passing the mobile phase, containing the mixture of the components, through the stationary phase, which consists of a column packed with solid particles. Physical and chemical forces acting between the solutes and the two phases are responsible for the retention of solutes on the chromatographic column. It is the differences in the magnitude of these forces that determine the resolution and hence separation of the individual solutes. Alternatively, the separation may be considered to be determined by the distribution of the solutes between the two phases.

The elementary forces acting on the molecules are of five types:

(1) London dispersion forces or van der Waals forces operate between molecules causing momentary distortion of their electrostatic configuration.

(2) Dipole interactions arise in molecules temporarily distorted and result in electrostatic attraction between the molecules.

(3) Hydrogen bonding interactions occur between proton donors and proton acceptors.

(4) Dielectric interactions resulting from electrostatic attraction between the solute molecules and a solvent of high dielectric constant.

(5) Electrostatic or coulombic interactions.

Any variable which affects these intermolecular forces will influence the degree of separation obtained by passage through the chromatographic column. It is difficult to develop a complete theory of chromatography since it is essentially a non-equilibrium process and the flow of solvent is generally too fast to allow equilibration of the solute molecules between the mobile and stationary phases. If this were not so, simple diffusion would be an important factor leading to band spreading and thus poor separation. In addition, the flow of solvent through a column is not easy to model since it depends on the shape of the stationary phase particles and the way in which they are packed together; furthermore the resultant eddies (see Section 2.3) produced in the solvent stream pose theoretical problems. The best we can hope to achieve is a theory which describes the average of the whole chromatographic column. For improved resolution one must aim for better equilibration rates by (*a*) either increasing molecular diffusion or reducing the distance molecules must diffuse and (*b*)

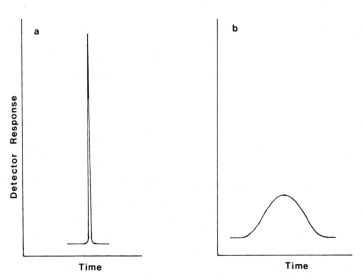

Fig. 2.1. Examples of an elution profile from a good (a) and a bad (b) chromatographic column.

improvements in the flow pattern by ensuring a uniform packing of the stationary phase.

The separation and purification of compounds by HPLC relies on a number of physical parameters which can be related to one another by theoretical considerations. To carry out successful HPLC separations it is important to have a reasonable understanding of these parameters and their interrelationship in order overcome problems which can arise in the practical use of HPLC. Firstly, to carry out efficient separations it is necessary to have an understanding of what constitutes the difference between a 'good' and a 'bad' column (Fig. 2.1). Elution of a component from a 'good' column occurs as a sharp, narrow band representing minimal dilution from the original sample concentration. Conversely, a 'bad' column will cause considerable dilution, resulting in the elution of a broad band and consequently little resolution between different components.

It is not only the stationary phase and its packing within the column that determines the quality of the overall chromatographic system. For example, the valves, tubing and detector may retain unnecessarily large volumes of the mobile phase; similarly the column end piece design may not generate an even liquid distribution; any of these parameters can cause considerable band broadening.

2.2. Basic terms

Before considering chromatographic theory in more detail it is important to understand the basic parameters. Fig. 2.2 shows a typical chromatogram, which represents the concentration of solute eluting from a column (as determined by the response of a sample detector) plotted against either time or volume.

After injection of the sample onto the column the compounds that do not interact with the stationary phase will be eluted at time t_0 in the void volume V_0. The void volume represents the sum of the interstitial volume between the particles of the stationary phase and the accessible volume within the particle pores. The retention time of the sample (t_r) is the time from injection to the time of maximum concentration in the eluted peak. The retention volume V_r is the volume of solvent required to elute the solute as measured from the

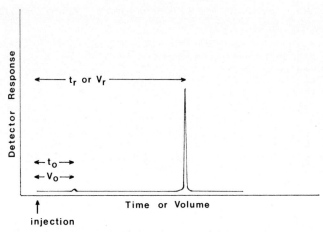

Fig. 2.2. A typical HPLC profile showing various parameters defined in the text.

centre of the chromatographic band. V_r and t_r can be related by the following equation:

$$V_r = t_r \times f \tag{2.1}$$

where f represents the flow rate.

The retention volume is directly related to the distribution or partition coefficient of the solute (K) between the stationary and mobile phases:

$$V_r = V_m + (K \times V_s) \tag{2.2}$$

where V_m is the volume of the mobile phase and V_s is the volume of the stationary phase.

The efficiency of a chromatographic column is measured by the number of theoretical plates (N) to which the column is equivalent. The term was originally used to describe the process of distillation and can be visualised as a series of hypothetical layers in which the solute concentrations in the relevant phases are assumed to be in

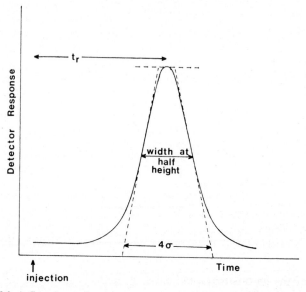

Fig. 2.3. A Gaussian peak; often assumed to be the idealized HPLC peak shape.

equilibrium. The number of theoretical plates is also a measure of the amount of band broadening caused by a column. It can be calculated from the formula:

$$N = (t_r/\sigma)^2 \tag{2.3}$$

where σ is the mathematical term representing the standard deviation of a Gaussian peak.

It is generally assumed that, initially, injection volumes will spread to give a Poisson distribution and then to a Gaussian distribution. A typical Gaussian peak is shown in Fig. 2.3.

Determination of plate number is simple from a detector chart recording, since tangents to such a curve at the point of inflexion intercept the baseline at a distance of 4σ apart. Assuming that peak shapes are Gaussian the peak width is therefore given by 4σ. This can be the parameter which is measured from the chromatogram and so the above formula can be written as:

$$N = 16 \times (t_r/4)^2 \qquad (2.4)$$

However, the method giving rise to the least error is to measure the peak width at half height. Using the mathematical properties of a Gaussian curve this gives rise to the formula:

$$N = 5.54 \times (t_r/\text{width at half height})^2 \qquad (2.5)$$

N is often quoted as a measure of the column performance and the larger the number, the better the column. In general, N increases for small particle diameters, low flow rates, higher temperatures, less viscous solvents, small solute molecules and good columns. The value of N is independent of the retention time of a solute but is proportional to column length. To enable a direct comparison to be made between different columns it is more useful to use the height equivalent to a theoretical plate, H (also referred to as the plate height). It is given by the formula:

$$H = L/N \qquad (2.6)$$

where L is the length of the column.

H compares the amount a solute has spread with the distance it has travelled and it follows that efficient columns have small values for H. It can be shown that the distance over which a solute is dispersed at the end of a column is given by:

$$4\sqrt{H \times L} \qquad (2.7)$$

Thus columns should also be kept short to minimise band dispersion. Since the plate height is a function of particle size it is sometimes useful to refer to the parameter known as the reduced plate height (h), which is a measure of the number of particles per plate. It can be determined from the equation:

$$h = H/d_p = L/N \times d_p \qquad (2.8)$$

where d_p is the diameter of the particles. H is dimensionless and is typically 2 for a well-packed column.

2.3. Band broadening

One of the undesirable features of liquid chromatography is the band broadening or dilution which occurs during passage of the solute through the chromatographic column. There are four important factors which contribute to band broadening, these are eddy diffusion, longitudinal diffusion, variations in mass transfer and extra-column effects

Eddy diffusion is caused by variable size channels through which the solute molecules pass. In wider channels the molecules travel faster than in narrow channels and therefore give rise to band broadening; the shape of the solid phase particles determines the actual effect of eddy diffusion on band broadening and can be held to a minimum by using spherical particles of uniform size and packing them evenly into the column. The degree of eddy diffusion is similar at all flow rates.

Longitudinal diffusion refers to diffusion of solute molecules in the direction of flow and, since solute diffusivity is low in liquids (providing that the analysis time is not excessively long), it is a negligible problem.

Mass transfer problems arise from limitations in the rate of diffusion of solute molecules between the stationary and mobile phases. In particular, molecules that diffuse more deeply into the stationary phase will lag behind those which diffuse less deeply. This will be accentuated by high flow rates and can be best minimised by coating the stationary phase thinly over a non-porous particle. However, such pellicular packings suffer from the disadvantage that there is a relatively large, chromatographically inactive core which reduces the loading capacity of the column.

Extra-column band broadening is caused by either excessive dead space within the chromatographic system (e.g. large internal volume fittings, tubing, detector cells) or alternatively by an inefficient method of sample introduction. It is therefore important to use low, or better

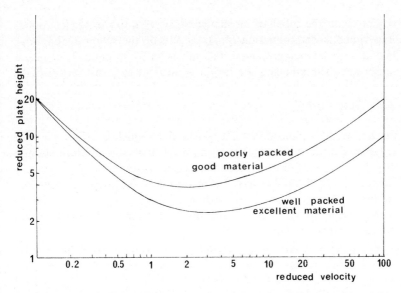

Fig. 2.4. A Knox plot. Reproduced from Bristow (1976) with permission.

still, zero dead volume fittings and to have the injection device as near as possible to the column.

These four factors cause band broadening and work additively such that their summation contributes to H, the height equivalent to a theoretical plate. This parameter is dependent on flow rate due to the contribution from longitudinal diffusion and mass transfer. Band broadening can therefore be minimised by use of an optimum flow rate. This relationship is usually visualised in a Knox plot (Fig. 2.4) using the reduced plate height (h) as the vertical axis and the reduced velocity (v) as the horizontal axis. The reduced velocity is given by the formula:

$$v = u \times \left(d_p / D_m \right) \qquad (2.9)$$

where u is the linear velocity, d_p is the diameter of the particles and D_m is the diffusion coefficient of the solute in the mobile phase.

The reduced velocity compares the linear flow velocity through the column with the speed of molecular diffusion into the particle pores. Experimental determination of D_m is outside the scope of this book but nevertheless experimental determination of optimum flow rate is important in minimising the height equivalent to a theoretical plate.

2.4. Retention

So far we have considered the physical parameters relating to the column and the stationary phase; however, the value of chromatography lies in the ability to separate components contained within a mixture and this ability can be quantitated. In order for any two solutes to be separated they must be differentially retained during their passage through the column so that different volumes of solvent are required to elute each solute. This is demonstrated in Fig. 2.5.

Compound A has been separated from compound B. A non-retained compound would be eluted in the void volume V_0 at time t_0. A

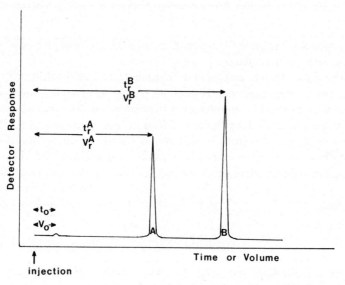

Fig. 2.5. Representation of the elution of two components A and B and their respective chromatographic parameters.

retained compound elutes after a greater volume V_r has passed through the column in a longer time t_r. A common measure of the degree of retention is the capacity factor (k'), which can be calculated from the following equation providing that the flow rate remains constant:

$$k' = (V_r - V_0)/V_0 = (t_r - t_0)/t_0 \qquad (2.10)$$

where k' is the number of column volumes required to elute a particular solute. Theoretically k' can be defined as the ratio of the mass of solute in the stationary phase to the mass of solute in the mobile phase. Using this definition, capacity factor (k') can be related to the fundamental parameter K, the distribution or partition coefficient:

$$k' = M_s/M_m = (C_s/C_m) \times (V_s/V_m) = K \times (V_s/V_m) \qquad (2.11)$$

where M_s is the mass of solute in the stationary phase, M_m is the mass of solute in the mobile phase, C_s in the concentration of solute in the stationary phase and C_m is the concentration of solute in the mobile phase. From this relationship it can be seen that solute retention is directly related to the thermodynamic distribution between the two phases. It follows that retention can be altered by a change in the chemical nature or temperature of the two phases or by changes in volume of the phases.

A parameter closely related to capacity factor but less commonly used is the retention ratio R, which is defined as either the fraction of solute in mobile phase or alternatively as the ratio of the distance travelled by the solute to the distance travelled by the solvent:

$$R = t_0/t_r = 1/(1 + k') \qquad (2.12)$$

R is normally used to express results from a thin layer chromatography plate or a paper chromatogram whereas k' is reserved for column chromatography.

The separation of component A from component B can be expressed by the separation factor or selectivity (α), which is simply the ratio of capacity factors:

$$\alpha = k'(A)/k'(B) \qquad (2.13)$$

2.5. Resolution

Ideally, a solution of a mixture of components applied to a chromatographic column in a small volume would be totally separated and each solute eluted in the same small volume. However, as discussed previously, solute bands broaden as they move down a chromatographic column. For a column of length L we have seen that the broadening increases proportionally to $\sqrt{L}$ whereas the separation of band centres is directionally proportional to L. Consequently, a sufficiently long column will always give a required separation, but in practice the solute will either be too dilute to detect, or the pressure needed to cause flow will be too high to permit safe operation of the equipment. Such problems can be overcome by a careful choice of stationary and mobile phases such that the solutes are selectively retained. The resolution (R_s) of two components is a measure of how well they are separated (Fig. 2.5) and is defined by the equation:

$$R_s = \text{difference in retention time/mean peak base width}$$

$$= 2\left(t_r^B - t_r^A\right)/\left(W^A + W^B\right) \tag{2.14}$$

where W^A and W^B are the peak base widths in the same units as t.

For two Gaussian shaped peaks if the resolution $R_s = 1$ the peaks will be effectively separated and have an overlap of only 2%. With a resolution of 1.5 or above there is excellent separation with no overlap. However if there is any distortion in peak shape then this will increase the possibility of overlap between adjacent peaks. In the above equation, W can be substituted by plate number (N) which then gives rise to an equation expressing the resolution in terms of three fundamental factors; the selectivity α, the capacity factor k' and the plate number N. Thus:

$$R_s = (1/4) \times \left[(\alpha - 1)/\alpha\right] \times \left[k'^B/(1 + k'^B)\right] \times N. \tag{2.15}$$

selectivity factor	retention factor	column efficiency factor

These three factors can be optimised independently. The selectivity factor is the most important in determining resolution and as previously defined it is the ratio of the capacity factors for the two components. Obviously α must not be equal to 1 or there will be no resolution. The selectivity can be varied by changing either the temperature (affecting diffusion processes) or the composition of the mobile or stationary phases. Optimisation of α is usually achieved by trial and error and the best method is to change the nature of the mobile phase which allows improvement in α with little change in the capacity factor. To change the stationary phase is impractical. Changing the temperature has a more pronounced effect in ion-exchange chromatography than in other chromatographic modes. Large changes in α can be achieved by secondary means such as chemical complexation.

The term $k'^{B}/(1 + k'^{B})$ is referred to as the degree of retention factor and needs to be considered mainly when k' is small. In situations where solutes are strongly retained, the term varies only to a small extent and eventually has a negligible influence on the resolution. Consequently there is little point in increasing retention to resolve two close peaks. Similarly, if k' is less than 1 the resolution decreases rapidly.

The plate number N can be increased by using a longer column, by decreasing the particle size or by optimising the flow rate (from the Knox plot). However, since resolution depends on $\sqrt{N}$, it can be impractical to significantly increase the resolution by means of this alone.

2.6. Separation time

It is of value to the chromatographer to observe the influence of changing various parameters on the separation time, so that an optimum separation can be achieved in minimum time (in addition, this approach will minimise solvent consumption). It can be shown that the time taken to elute the last peak is given by:

$$t_{R} = 16 \times R_{s}^{2} \times [\alpha/(\alpha - 1)] \times \left[(1 + k')^{3}/k'^{2}\right] \times [H/u] \quad (2.16)$$

In this form the equation shows that whilst other terms are constant, doubling the resolution causes a four-fold increase in separation time. The equation also shows that the separation time is shortest when k' is 1.26 but for practical purposes the range 1.5–5.0 gives the optimum resolution time. For a complex mixture of solutes it is generally not possible to achieve good resolution between all components with a maximum k' value of 5. Typically k' values of up to 10 offer a practical solution. Since the term containing the capacity factors plays a significant part in this equation, the spread of k' values can be minimised by using either gradient elution or by coupled columns.

2.7. Preparative separations

Scaling-up an analytical separation allows the isolation and purification of a particular component from a mixture of solutes. Preparative separations can be achieved without reducing column performance parameters if care is taken with establishing chromatographic conditions. Preparative HPLC simply describes any chromatographic separation in which the eluted components are collected, but more usually refers to separations in which higher sample loadings or larger columns are used. Columns used for preparative scale separations have been classified according to size (Verzele and Geeraert, 1980). For example, columns up to 25×0.46 cm have been classified as the PLC-1 category and this group includes analytical size columns in which extra large samples may be applied (up to 3 mg). Slightly larger columns (50×0.6–1.0 cm) which can still be handled by analytical instruments have been classified as PLC-2 and their capacity has been estimated to be up to 25 mg depending on the separation. Larger columns have been classified as PLC-3 and PLC-4 depending on whether they are long and narrow or short and wide and will typically separate quantities up to 10 g. In analytical HPLC a particular stationary and mobile phase will give a fixed plate number (N), retention (k') and resolution (R_s) for a particular sample with a given flow rate and temperature. For preparative purposes, simply increasing the sample concentration will eventually have serious effects on

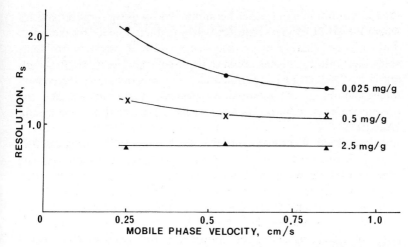

Fig. 2.6. Effect of mobile phase velocity and solute weight on resolution. The plots show milligrams of diethylketone injected per gram of packing. Chromatographic conditions: column, Porasil A (50×1.09 cm I.D.); Mobile phase, chloroform (50% saturated with water); flow rate, 0.25 cm/s; temperature, ambient. Reproduced from DeStefano and Kirkland (1975), with permission.

these parameters by upsetting the equilibrium established in the column. Fig. 2.6 shows the resolution of a column as a function of mobile phase velocity and solute weight. Resolution is decreased by increasing the loading; this arises as a result of decreases in the values of α and k'. The column is considered to be in an overload condition when k' values for sample peaks show a greater than 10% decrease. At higher loadings the flow rate has little effect on resolution (providing that the flow rate is linear and no turbulence is introduced). This is an important observation because it means that flow rate can be increased to improve the yield per unit time. This latter factor, also called the production rate, is an important consideration when optimising the conditions under which the separation is carried out. From the classical Knox plot it has already been shown that in a non-overloaded condition the mass transfer term contributing to band broadening increases with increasing flow rate. Consequently, for preparative separations where production rate is important, the effect of flow rate on resolution can further be minimised by using similar particles where the effect of mass transfer is less marked. Obviously there is a limit to the reduction in particle size because of practical problems associated with the higher pressure requirements. The maximum loading is best determined experimentally and will obviously depend on the resolution of the required components. Equations relating the production rate to basic parameters have been derived (Hupe and Lauer, 1981).

In the practical development of a good preparative separation it is a good idea to start with a well-researched analytical separation. The transfer to a preparative column can then take place which often involves reducing the solvent strength to allow capacity factors (k') of up to 10 and which thereby improves sample loading since concentrations of the solute along the column will be reduced. This also has the effect of increasing resolution. Introduction of gradient elution will facilitate recovery of components with high capacity factors. Since much larger samples can be applied for separations with well-resolved peaks it may be of value to further investigate improvements in the selectivity (α) by modification of the solvent system. A compromise between sample loading and mobile phase velocity can then be found to maximise production rate. It should be noted that sample loading can be increased in two ways: (a) by increasing the concentration and (b) by increasing the volume; the latter is considered to be preferable (De Stefano and Beachell, 1972). The production rate can then be further increased only by increasing the column dimensions. Wider columns are recommended for this purpose and production rate is directly proportional to cross-sectional area. Optimising the production rate not only optimises the system with respect to time but also to solvent consumption and to concentration of the sample in the eluent.

A useful review of preparative liquid chromatography can be found in the literature (Verzele and Geeraert, 1980).

CHAPTER 3

Instrumentation

3.1. Introduction

The rapid growth in the number of researchers interested in HPLC has been accompanied by a similar growth in the number of manufacturers of HPLC equipment. Consequently those who are new to the field of HPLC are often overwhelmed by the information which is available on the equipment. This chapter is intended to serve as a brief introduction to the workings of a fully operational chromatographic system to (*a*) enable those users who are perhaps not getting the best from their investment to improve on their system and (*b*) to serve as a guide for those who are thinking of investing in HPLC equipment. A major consideration is the cost. The simple isocratic system shown schematically in Fig. 3.1. can cost in the region of £6,000–£10,000 while for a fully integrated microprocessor controlled gradient system the cost may be anything up to £35,000. Clearly this is a substantial investment, particularly for grant-supported research groups and it is important that the initial choice of equipment is the correct one. For example, it may be wise to choose a system to which other components can be added later (modular design). Throughout this chapter items of equipment from various manufacturers have been cited. This reflects items with which the authors are familiar and is not meant as a specific recommendation.

3.2. System design

The flow diagram in Fig. 3.1 will be familiar to anyone using low pressure liquid chromatographic systems with the exception of the pressure monitor; however, each user will have their own specific requirements. For example, a user who wants a machine as a general all-purpose tool on which a multitude of tasks can be performed (thereby exposing the equipment to a variety of aqueous and non-aqueous solvents) will perhaps not require the same degree of sophistication as someone requiring a dedicated instrument, e.g. for quality control purposes.

Each of the components detailed in Fig. 3.1. will be considered in turn to try to create an understanding of how to design a system. A detailed description of the development of HPLC equipment can be found in McNair (1984).

3.3. Reservoir

All tubing and fittings should be chemically inert to corrosion and therefore most are made of either 316 grade stainless steel or some form of PTFE tubing. When using microbore tubing it is of paramount importance that no particulate material should be present either in the solvent or in the sample; filtration of both through a 0.45 μm filter (e.g. Millipore) should therefore be a routine practice and filters are available which can handle sample volumes as low as 5–50 μl. It is also a common practice to insert some form of in-line filter into the reservoir as an added precaution. Further details of solvent filtration and degassing are discussed in Chapter 10.

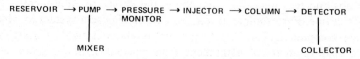

Fig. 3.1. Representative flow diagram of a simple isocratic chromatographic system, highlighting the individual components.

3.4. Pumps

The most critical feature of a liquid chromatographic system (apart from the column) is the pump. The ability to deliver solvent at a constant and precise flow rate, over long and short periods, has been the goal of manufacturers for the last decade. More than one approach has been taken. An early design was the gas pressure-driven syringe-type solvent delivery system known as the constant pressure syringe pump. The main disadvantage of this design was that the delivery was sensitive to pressure changes caused by temperature fluctuations or solvent viscosity changes. Another version of the syringe pump incorporated a mechanically driven piston which was insensitive to pressure change. The second disadvantage of the syringe-type pumps was the requirement for a large volume of solvent to be contained in the syringe chamber which could not be easily replaced during solvent changeover; moreover, the plunger had to sweep the length of the chamber to commence the next delivery stroke; this can cause serious baseline problems. (The use of twin-headed syringe pumps operating 180° out of phase has been incorporated into the FPLC (fast protein liquid chromatography) system introduced by Pharmacia to overcome this latter problem.) In general, however, syringe pumps and gas displacement pumps have largely been superceded by reciprocating pumps. The characteristics of each type of pump have been discussed elsewhere (Snyder and Kirkland, 1979). Most common types of reciprocating pumps operate a glass piston, via a cam, to displace a volume of liquid, the direction of which is controlled by check valves (Fig. 3.2). The simplicity of design and the reliability of performance has led to considerable effort being devoted to improvements in the performance of this type of pump at the expense of other designs. The major fault in this particular design is that, as the piston sweeps the pump chamber, a pulse of solvent is delivered which can cause uneven solvent flow and some form of dampening is required. The incorporation of hydro-pneumatic dampeners consisting simply of a length of flattened tubing introduces an undesired extra dead volume between the pump and the column; however, they are still incorporated in some pump designs. The introduction of a fast refill cycle in the piston reduces pulsation

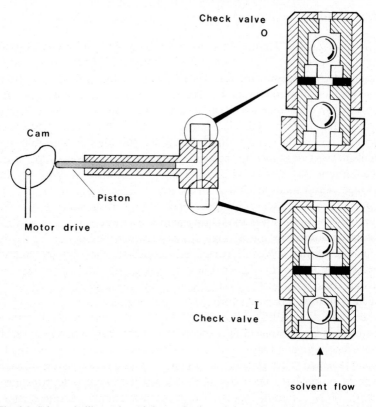

Fig. 3.2. Schematic illustration of the mechanism of the reciprocating pump. A piston, which is driven by a cam, alternatively draws in solvent through the inlet check valve (I) during the refill stroke and delivers a pulse in the forward stroke through the outlet check valve (O). The check valves are enlarged to show detail. The figure is drawn to illustrate the piston approaching the maximum forward position and therefore the outlet check valve is open and the inlet valve is closed. This allows solvent flow in one direction only.

dramatically and, in conjunction with a dampening device, produces relatively smooth solvent delivery (e.g. Altex 110 model). The introduction of a twin-headed pump whose pistons operate 180° out of phase produces essentially pulse-free solvent delivery since the refill stroke of one piston is compensated for by the fill stroke of the second piston (e.g. Waters M45 type). An extension of this is a triple-headed pump whose pistons operate 120° out of phase (e.g. Du Pont). Single-headed pumps in which a mechanically driven inlet valve is used to direct solvent flow (e.g. Varian 5000) and electronic controlled systems for synchronous operation of pistons to ensure pulse-free flow (e.g. Micrometics) are also available. More recently a pump with reciprocating piston speeds of 23 strokes per second has been introduced. At this stroke rate, the pulsations are too fast to be recorded by detectors and the solvent delivery is pulse-free (A.C.S. model 400). The more sophisticated models also have in-built mechanisms for compensating for the compressibility of different solvents which is an important feature for gradient delivery in reversed phase chromatography. In the isocratic mode, any of the above-mentioned types of pump can be incorporated into the simple system shown in Fig. 3.1. It is unlikely, however, that isocratic separations will suffice for all conditions and therefore anyone investing in HPLC should also consider a gradient system; for a *relatively* small increase in cost, a sophisticated programmable gradient controller can be purchased, which could also be used in the isocratic mode.

3.4.1. Gradient systems

Binary gradients are created by the selected mixing of two separate solvents. On a single-head two-pump system, by altering the flow rate of each pump, the homogeneous mobile phase composition is achieved using a low-volume stirred mixing chamber on the high pressure side of the pump (usually a stirring magnet with an internal filter). Reciprocating pumps are less precise at very low flow rates (0.1 ml/min) and therefore the initial and final stages of the gradient are less accurate. This may be important in the elution of some species whose capacity factor is critically dependent on mobile phase composition, and therefore does constitute a slight disadvantage in such systems. In addition, the accuracy of the gradient relies on the high performance of two separate pumps.

Those systems which utilise single multiheaded pumps also have the option of using low pressure mixing devices. These operate by altering the solvent flow through switching valves which are in turn controlled by solenoids. The use of solenoids means that more than two solvents can be precisely delivered, allowing ternary gradients to be easily made, and a wash cycle can be introduced. Since solvents are supplied continuously only reciprocating pumps can be used. The main advantage of low pressure mixing is the cost, since only one pump is necessary. Also, because the solvents are mixed before the high pressure side of the system, many of the problems associated with solvent degassing are eliminated. Alternative systems are available which give reliable and accurate gradient formation through sophisticated microprocessor control (Micrometrics).

3.5. Injectors

The early design of HPLC injector systems was strongly influenced by those used in gas chromatography (GC). This is obvious when one considers that, although septum injection is still popular in GC, it has been almost totally supplanted in HPLC by high pressure injection

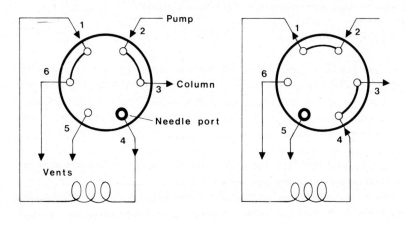

Load **Inject**

Fig. 3.3. Illustration of the arrangement of a high pressure injection valve. In the load position the sample is introduced via the sample loop (positions 1,4). Upon injection the solvent flow from the pump (2) is directed through the sample loop to the column (3).

valves of the type illustrated in Fig. 3.3. This is the most popular design of injection valve (e.g. Rheodyne, Valco) and permits zero dead volume injection at pressures up to 5,000 psi. Any alternative method, e.g. stopped flow septum injection, has been found inferior due to the generation of particulate matter from the septum after repeated injections.

3.6. Corrosion

The majority of HPLC hardware is made of either corrosion-resistant stainless steel (type 316), or other relatively inert materials (e.g. PTFE) where possible. The use of stainless steel fittings arose from the increase in column back pressure at high mobile phase flow rates, coupled with the use of low volume internal diameter tubing and fittings. At the present time many successful separations of biomolecules are carried out using this type of equipment in both reversed phase and ion-exchange modes (CRC Handbook of Chemistry and Physics, 1984, Vol. II). However, the use of halide ions in low pH mobile phases is not recommended by any manufacturer as corrosion of the stainless steel may occur and therefore certain precautions must be observed. Routinely, it is good practice to leave the HPLC system flushing with distilled water overnight. At the first sign of any corrosion, the system should be passivated with 20% nitric acid (flow rate, 1 ml/min for 30–60 min). These simple guidelines should prevent any major problems associated with the use of halide ions.

3.6.1. Proteins: a special case

Many globular proteins contain prosthetic groups with metal ions and sensitive thiol groups which may be 'poisoned' by slow leakage of metal ions from the welding in the joints of stainless steel equipment. To overcome this potential problem, two manufacturers have designed equipment to minimise contact with any stainless steel.

LKB manufacture a reciprocating twin-pump system with pump heads made of KEL-F. Gradients are formed by a low volume mixer which is located on the low pressure side of the pump, thereby eliminating some of the solvent degassing problems. The rest of the system is similar to most other HPLC systems, but has the added advantage that the pump heads can be interchanged with stainless steel heads for other applications (LKB Ultrochrom). Furthermore, low pressure chromatography is compatible with this equipment.

Pharmacia have introduced a syringe-type two-headed pump designed to withstand pressures up to 1,000 psi. The use of this pump design coupled with a switching valve offsets the requirement for check valves and in general gives a smooth pulse-free flow. Mixing of mobile phases for gradient HPLC takes place on the 'pressurised' side of the pumps. The system is compatible with a specialised range of columns introduced for protein separations with recommended operating pressures of less than 500 psi at 1 ml/min. This system can also be used for chromatography using conventional low pressure columns (Pharmacia FPLC system). A recent addition to the range is a reciprocating pump capable of operation at higher pressures and therefore suitable for use with other HPLC columns. A disadvantage of syringe pumps is that the volume of solvent in the syringe chamber is very large and must be totally displaced when changing solvent conditions to prevent carry-over of solvent to the next separation. However, with the introduction of microprocessor control a fast changeover can be achieved in a few minutes. A very useful feature of the FPLC system is the so-called 'super loop' which allows the loading of large sample volumes (up to 50 ml) onto the column without the introduction of air. A similar piece of equipment for conventional HPLC apparatus would be most useful since at present the only way to load a large volume is through either the pump heads or through very large sample loops.

Applied Biosystems Ltd. recently made an addition to the list of HPLC equipment devoted to proteins. The range consists of a protein sequencer (470 A), a protein purification system (130 A), a peptide synthesiser (430 A) and an amino acid analysis system (120 A). The chromatographic module is a microbore HPLC system specifically designed for use with very small quantities of material. The columns consist of Brownlee cartridges. Flow rates in the order of 10 to 100

μl/min are produced by sophisticated microprocessor-controlled syringe pumps (ABI 130). This equipment is expensive but is of the highest quality.

3.7. Detectors

This section is not written as a detailed description of each category of detector but as an overview to allow the reader to gain a basic understanding of the potential uses of a given detector. For a fuller account of the use and operation of detectors the reader is referred elsewhere (Scott, 1977). Detection can be conveniently categorised into those modes which depend upon:
(1) Some property of the solute of interest.
(2) Differential behaviour of the solute and solvent in respect of some general property.
(3) Removal of the solvent to detect the solute independently.

3.7.1. Ultra violet (UV) detection

Without doubt UV detectors are the most popular detectors currently used in combination with HPLC and are favoured because of their versatility, reliability, sensitivity and relatively low cost. Detection depends on the presence of a suitable chromophore and is completely non-destructive. The relationship between sample concentration and absorbance can be defined by the Beer-Lambert equation:

$$\log(I_0/I) = E \times L \times C \qquad (3.1)$$

where I is the transmitted light, I_0 is the incident light, E is the molar extinction coefficient (at that particular wavelength), L is the optical path length and C is the molar concentration. The simplest and most inexpensive type of detector is known as a fixed wavelength detector (FWD). These detectors generally use a mercury lamp as a UV source and utilise different interference filters to enable selection of the wavelength of interest. For example, some of the popular filters are 214, 220, 254 and 280 nm. Alternative light sources are available for monitoring at low wavelength: zinc at 214 nm and cadmium at

TABLE 3.1
Operational wavelengths for various light sources

Source	Wavelength (nm)
Mercury	254
	313
	365
	405
	436
	546
	578
Cadmium	229
	326
Zinc	214
	308
Magnesium	206
Deuterium	190–350 (continuous)

229 nm (Table 3.1). For routine operations, such as in quality control monitoring, a simple UV photometer is usually adequate; however, for research work where compounds may need to be monitored at various wavelengths an alternative is to use a variable wavelength UV detector (VWD). These detectors are of two types; firstly, those which use a visible or UV source (tungsten lamp and deuterium lamp, respectively) in combination with a monochromator to select the desired wavelength, and secondly, those which use a photodiode array to allow monitoring either at a fixed number of wavelengths or across a whole spectrum of wavelengths. Photodiode array detectors have become increasingly popular over the last five years because of their ability to monitor the entire absorbance spectrum of an eluting peak and thereby 'fingerprint' that peak as an aid to peak identification. In combination with sophisticated computer graphics a detailed spectrum of eluting peaks can be obtained. Photodiode array detectors have a number of advantages over the alternative motor-driven monochromator UV detectors which rapidly scan eluting peaks. Two major advantages of the photodiode array detectors are that they possess less moving parts (thus reducing the risk of mechanical failure) and

also that they are less prone to peak skewing at high flow rates.

When monitoring at low wavelengths care should be taken in the selection of the solvents used in the mobile phase as several popular solvents have UV cut-offs at relatively high wavelengths.

It should be emphasised that generally HPLC-grade solvents should be used, especially where detectors are being used at high sensitivity. Moreover, the use of purified solvents may help to extend the lifetime of analytical columns (Chapter 10).

The sensitivity of UV detectors is generally in the low nanogram range although this can be increased by sample derivatisation. The sensitivity of detection may also be increased by alterations in the design of the detector flow cell and it is important to emphasise that the volume of the flow cell should be chosen to match the HPLC system; thus a very low volume flow cell (e.g. 1–2 μl) should be used with microbore, or very small particle columns where very high column efficiencies are anticipated, whereas larger flow cells (e.g. 5 μl) should be used with more conventional systems.

3.7.2. Fluorescence detection

Fluorescence detection is often used where no other property of the solute (e.g. UV of RI detection) is convenient and can be either an intrinsic property of the solute itself or a derivatised form of the solute. Solution studies have indicated that the sensitivity of detection can be increased by up to three orders of magnitude over UV. This has increased the popularity of post-column fluorescence detection methods for may compounds, including physiological fluids, catecholamines, and other polyamines. A popular use of fluorescence detection is in peptide chemistry where no convenient intrinsic chromophore is present. Derivatising agents such as orthophthalaldehyde and fluorescamine are used extensively in both pre- and post-column systems allowing detection of low picomole quantities (Chapter 11). In addition, detection can be performed using the intrinsic fluorescence of many compounds such as steroids, vitamins, and nucleotides.

The conventional fluorescence detector consists of the following components:
(1) A light source, either xenon or mercury arc, or quartz halogen.

(2) A wavelength selector using either a filter or monochromator. A series of filters are available between 254 nm and 520 nm (excitation) and 420 nm and 680 nm (emission). Light is emitted in every direction from the sample but is usually monitored at right angles to the excitation radiation. This can mean that a lower sensitivity is achieved since less light is seen; however, if a parabolic reflector is incorporated inside the cuvette chamber more of the light can be collected (e.g. Kratos; Schoeffel, Sf 970).
(3) A photomultiplier; this provides good sensitivity over a wide range of wavelengths.
(4) A sample cell, the size of which may vary between 1 and 40 μl.

Further innovations have been the introduction of a laser light source which is more monochromatic than conventional light sources and therefore does not suffer interference from extraneous light.

The increased sensitivity of fluorescence detectors means that the presence of contaminating material in the mobile phase can become a major problem. One method of circumventing this is to distill the solvents over a fluorescent reagent prior to use (Stein and Brink, 1980).

3.7.3. Refractive index (RI) detection

Refractive index detection is the second most popular non-destructive method used in HPLC. RI detection utilises the principle of the change in direction of a beam of light as it passes through different matrices. Different methods have been used to monitor this change.

3.7.3.1. Fresnel prism.
This method measures the change in the amount of light transmitted through the surface of a glass prism and the mobile phase. Two parallel beams of light are focussed through the prism onto the detector. The light which is transmitted through the interface is scattered, and a collecting lens is used to focus the reflected light onto a photocell. When the solute passes through the sample cell the light beam is refracted differentially. A tungsten lamp is used as a light source. This type of instrument is supplied by Laboratory Data Control.

3.7.3.2. Deflection refractometer.   This method measures the refraction of a parallel beam of light as it crosses the interface of two media. A light beam from a tungsten lamp is passed through the flow cells and is reflected back and focussed onto a photocell. One of the flow cells is the reference cell, the other being the sample. When the solute enters the sample cell the light is deflected. This type of instrument is supplied by Waters (R 400).

3.7.3.3. Interferometric refractometer.   This method measures the difference in the speed of passage of two beams of light between the sample and reference by the interference of the two beams after passage through the cells. The light beams pass through the flow cells and are recombined and collected in a photomultiplier. As the relative speed of light between the sample and reference changes the beams

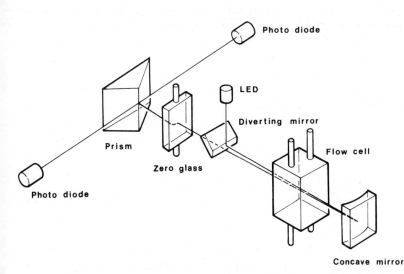

Fig. 3.4. Schematic illustration of a refractive index detector (LKB 2142-010). The LED light source (950 nm emission) is directed through the flow cell and reflected back by a concave mirror. The light is split into two beams at the edge of a prism and directed towards two photodiode sensors. Reproduced from LKB technical brochure with permission.

become asynchronous, interference of the two beam becomes destructive and the energy reaching the photomultiplier decreases (Optilabs).

Recent additions to these categories incorporate some of the features of both the Fresnel optics and the deflection refractometer (LKB). The arrangement is shown in Fig. 3.4. Refractive index monitors have a major drawback in that they cannot be used with gradient chromatography. In addition baseline drift due to temperature and pressure changes are also a problem; however, for isocratic separations (particularly in size exclusion chromatography) the method is simple and generally trouble free. Sensitivity, however, is in the microgram range.

3.7.4. Radioactivity monitors

Radiolabelled compounds are frequently used to analyse metabolic pathways, drug pharmacokinetics and pharmacodynamics, and are also used as internal standards to quantitate extraction efficiencies. Until the last decade, these radiolabelled compounds were routinely separated by thin layer chromatography, paper chromatography, or column chromatography prior to tedious and time-consuming quantitation techniques, e.g. scraping, elution, etc. The advent of HPLC enabled rapid separations of radiolabelled compounds but then required sample aliquots to be collected with a fraction collector followed by counting on a scintillation spectrometer, a process which was both very expensive and time consuming when a large number of compounds in a sample had to be analysed. Subsequently, continuous flow radioactivity monitors have been developed.

Most radioactivity monitors use flow cells of different types positioned between two photomultiplier tubes which usually possess special reflectors to ensure high continuing efficiencies. The signal pulses from the photomultipliers are measured in coincidence to suppress contributions from noise and chemiluminescence. The signals from the photomultiplier tubes are amplified and are then usually subjected to various data reduction processes.

The flow cells are available in three forms:

Heterogeneous. The flow cells contain solid scintillator granules through which the eluent passes in exactly the same manner as through the flow cell of UV detector. The radioactivity detector may be positioned either before or after a UV detector when used with this flow cell.

Homogeneous. The eluate is mixed with liquid scintillant before passing through the flow cell. The sample is passed to waste after counting.

High energy. The eluent is isolated from scintillator material, but the flow cell is surrounded by granular or liquid scintillator.

A variety of features are associated with the equipment from different manufacturers which may provide useful options. Several manufacturers offer a splitter–mixer facility which allows a portion of the eluate from a column to be split off and taken to a fraction collector. The remainder of the eluate from the column is then automatically mixed with a metered volume of scintillant before passing through the flow cell.

A limitation of some machines is their inability to deal with the altering quench characteristics of the mobile phase in gradient HPLC systems. Some radioactivity detectors compensate for this by allowing a standard curve to be stored which allows DPM values to be automatically calculated, giving accurate peak determinations in a gradient chromatographic run. Similarly, this facility allows mixed isotope monitoring by determining the spill-over of the higher energy isotope into the counting window of the lower energy isotope. Several manufacturers now market machines which are linked to microcom-

TABLE 2

Counting efficiency and sensitivity of detection in radioactivity monitors

Flow cell	Counting efficiency		Approximate sensitivity of detection	
	^{14}C	^3H	^{14}C	^3H
Heterogeneous	30%	> 3% [b]	80 dpm	400 dpm
Homogeneous	55%	50%	100 dpm	250 dpm
High energy	55%	50%	n.a. [a]	n.a.

[a] Data not available.
[b] Dependent on scintillant.

puters that allow detailed data collection and processing. These facilities are extremely useful when monitoring unknown samples which may have a number of low energy peaks.

The counting efficiencies and sensitivity of detection of radioactivity monitors vary with the individual machine, the type of flow cell and the isotope. However, typical specifications for a good machine are shown in Table 3.2.

3.7.5. Electrochemical detectors

Electrochemical detection in combination with HPLC has become increasingly popular over the last decade for the measurement of easily oxidisable and reducible organic compounds. This rapid expansion of interest was triggered by the recognition of its use in the quantitation of biogenic amines from the central nervous system. More widespread use was fostered by the inherent advantages of the technique, these being selectivity, high sensitivity and low cost.

Electrochemical detectors are of two classes:

(*a*) Bulk property detectors which measure changes in an electrochemical property of the bulk fluid flowing through the detector cell. The most popular detectors of this class are conductivity detectors which apply a potential across the cell so that ions in solution will move towards the attractive electrode with a consequent change in conductivity. Two major disadvantages of conductivity detectors are that the use of buffer salts in the mobile phase should generally be avoided and also that the system is temperature sensitive. Nevertheless, these detectors have found some use in the quantitation of both organic and inorganic ions.

(*b*) Solute property detectors which monitor the change in potential or current as a solute passes through the detector cell. The more popular of these detectors are either coulometric or amperometric of which the latter are at present the more widely used. The detailed theory of these detectors is beyond the scope of this book and the interested reader is referred elsewhere (Kissinger, 1983; Stulik and Pacakova, 1982). Briefly, however, if a solute passes over an electrode which is held at a constant potential which is sufficiently high for an electron transfer and consequent oxidation or reduction to occur, then a current will be produced which is proportional to the solute concentration. In amperometric detectors the concentration of the solute entering and leaving the detector is the same (in practice up to 5% may be converted to product). In coulometric detectors the solute is totally converted (usually approximately 95% conversion). While the signal given by coulometric detectors is obviously greater, this advantage is often outweighed by the increase in background current and consequent decrease in signal to noise ratio.

The working electrodes used with amperometric detectors are dependent upon both the mode of HPLC and whether oxidative or reductive electrochemical detection is used. In the oxidative mode the working electrode is usually carbon paste or glassy carbon. Carbon paste is a mixture of graphite and an inert binding material such as paraffin oil, silicon grease or wax. These electrodes are easy to prepare, possess low background currents and are reasonably reproducible. Unfortunately, their use is limited to mobile phases where the organic modifier is at a low concentration (e.g. less than 25% (v/v) methanol or 5% (v/v) acetonitrile). Glassy carbon is a chemically resistant, mechanically rigid, glassy material. The sensitivity of glassy carbon and carbon paste electrodes is similar but the latter sometimes possess lower background currents.

The most popular of the electrodes used in reductive electrochemical detection utilise an electrode consisting of mercury deposited as a thin film on a gold substrate. This design overcomes the problems of vibration and mechanical instability observed with the older design of mercury drop electrodes.

A major limitation with electrochemical detection is the restricted choice of mobile phase since high concentrations of non-polar organic solvents cannot be used because of their inability to support conductivity. When buffers are used in the mobile phase they are usually in the 0.01–0.1M range to provide conductivity while maintaining a low background current. It should be emphasised that the constituents of the mobile phase should always be of the highest purities to prevent excessive background currents. The most popular detector cells used in electrochemical detection are thin layer and may be used with either amperometric or coulometric detectors. The most popular design (Bioanalytical Systems) has the mobile phase flow parallel to the electrode surface and a planar auxiliary electrode placed opposite the working electrode with the thin layer channel functioning as a salt bridge to the reference electrode. An alternative design is the wall-jet detector where the flow is perpendicular to the electrode surface (Fleet and Little, 1974); however, although this cell is designed to provide more effective mass transfer to the electrode surface, conventional cells have generally demonstrated superior performances.

The majority of applications with electrochemical detectors utilise a single electrode; however, it is relatively simple to devise a multiple electrode system which utilises two working electrodes in parallel with a corresponding auxiliary electrode positioned opposite and a reference electrode downstream. This arrangement allows current to be monitored at two potentials and is therefore analogous to monitoring at two different wavelengths with UV detectors. This method can be used to aid in the identification of an eluting solute and may also be used to quantitate an easily oxidised solute in the presence of substances which react at higher potentials, thereby conferring added sensitivity of detection.

A thorough review of the applications of electrochemical detection is beyond the scope of this book. Briefly, the classes of compounds that have been most frequently investigated include; aromatic amines, phenols, thiols, nitro-compounds, quinones, phenothiazines, purines and semicarbazides. Additionally, both ascorbic acid and uric acid can be conveniently monitored using electrochemical detection.

A number of hints specifically applicable to electrochemical detectors can be given as guidelines to obtain sensitive detection:

(a) Ensure that the electrode is compatible with the mobile phase and that the electrode surfaces are in good condition.

(b) Operate at the minimum potential required to oxidise or reduce the solute of interest.

(c) Ensure that the detector is electrically isolated and that all components are grounded to the same common ground connection.

(d) Minimise temperature fluctuations around the detector cell.

(e) Check that no air bubbles are present in the flow cell.

(f) Ensure that all constituents of the mobile phase are of the highest quality.

(g) Passivate the HPLC system regularly.

3.8. Fraction collectors

A variety of fraction collectors are available from a number of manufacturers and include those which are able to collect low micro-litre volumes in addition to those used in preparative HPLC which can collect much larger volumes. Fraction collectors which are to be used with HPLC systems require very rapid tube changing facilities and should also be resistant to the solvents used with HPLC systems; thus, fraction collectors which are designed for use with low pressure chromatography are usually inadequate. A major feature in the more recently marketed fraction collectors has been the introduction of advanced microprocessor control which provides a number of capabilities ranging from simple timed collection to total integrated control of the whole HPLC system.

Both LKB and Pharmacia supply rotary fraction collectors which are ideal for general purpose use in HPLC systems and are also relatively compact. The Pharmacia fraction collector utilises a pressurised pad on a contact arm to change collection tubes, while the LKB collector has a fixed arm with a mobile carousel which effects tube changeover. Both the LKB and Pharmacia machines are relatively inexpensive. A more sophisticated (and consequently more expensive) range of machines is manufactured by Gilson which allow a variety of tasks to be performed through microprocessor control. This machine allows either selected peaks or time-windows to be collected and the sample volume which can be collected is virtually unlimited. Moreover, clever programming and slight mechanical modification of this machine enables it to derivatise samples in a pre-column mode (e.g. with OPA) and then automatically inject derivatised samples.

3.9. Automation

3.9.1. Autosampler

The most obvious step in the automation of a chromatographic system is to automate sample injection. The ability to perform repeated functions without operator attendance is a labour-saving method available to all chromatographers. Machines such as the Waters Intelligent Sample Processor (WISP) and Kontron MS1, which are accurate down to less than 10 μl with negligible cross-contamination, are suitable for most applications. Alternatively the Gilson machine previously described (Section 3.8) is also worth considering.

3.9.2. Microprocessor control

At every stage of the chromatographic process illustrated in Fig. 3.1, some form of integrated process control can be inserted, as illustrated in Fig. 3.5. This may be of two forms in current HPLC machines. Firstly, the more popular machines have built-in microprocessor control modules with specially written software, usually on floppy disc or tape drive (e.g. Waters, Varian, Hewlett Packard, Altex, Pye Unicam, Micrometrics). This system of control maintains the instrument as one specifically devoted to the task. An alternative form uses a microcomputer (e.g. Apple II) with specially written software (e.g. Gilson, ACS, LKB). This latter system has the added advantage that the microcomputer can be used for other purposes when not being used to control the HPLC machine. These systems also offer a peak integration package, in addition to general system control.

A comprehensive list of manufacturers and the type of components currently being used in HPLC can be found in McNair (1984).

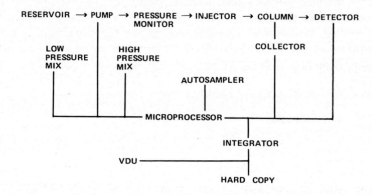

Fig. 3.5. The impact of microprocessor control on laboratory automation.

CHAPTER 4

High performance ion-exchange chromatography

4.1. Introduction

Ion-exchange chromatography was the first mode of chromatography to be developed to the point that we now term high performance liquid chromatography. The advances began in the early 1950s with a requirement to improve amino acid analysis techniques. The pressures used were initially quite low (about 100 psi, the working limit of glass columns). The same requirement was later demanded for the analysis of many other biochemical compounds, such as nucleotides in urine and blood. The chromatographic supports were based on spherical beads of polystyrene–divinylbenzene copolymer resins which were gradually made with smaller and smaller diameters with the aim of improving performance. The advent of precision bore stainless steel columns meant that greater demands were placed on the resins and cross-linked polystyrene was found to be too soft to withstand the higher pressures. This led to a search for ion-exchange column packings that could not only withstand the higher pressures but were also amenable to microparticle packing technology. Pellicular packings were developed which consisted of glass beads with ionic molecules attached in a layer on the surface. Pellicular packings were well suited to the high pressures but the beads were relatively large and gave a low chromatographic efficiency. They also possessed a relatively low capacity when compared with the original polystyrene–divinylbenzene resin supports. However, through the 1970s, silica-based ion-exchange resins were developed (Knox and Pryde, 1975) and sample capacities have now improved to levels which provide acceptable chromatography, although modern amino acid autoanalysers still use cross-linked polystyrene stationary phases.

4.2. Mechanism

Ion-exchange chromatography depends on the interaction between charged molecules on the stationary phase and the charged solutes and ions in the mobile phase. In simple anion-exchange chromatography, solute ions Y^- compete with mobile phase ions X^- for positive sites ($+$) on the stationary phase (these are often tertiary ammonium moieties such as DEAE) (Fig. 4.1). Alternatively, in cation-exchange chromatography, solute ions Y^+ compete with mobile phase ions X^+ for negative sites ($-$) on the stationary phase (these are often sulphonic acid moieties such as sulpho-propyl or carboxymethyl). Separation is achieved because sample ions that interact weakly with the ion-exchange stationary phase will be weakly retained on the column and elute early whereas solutes that interact strongly with the ion-exchange stationary phase will be strongly retained and will elute later. These effects are shown in Fig. 4.2 where there is an increase in

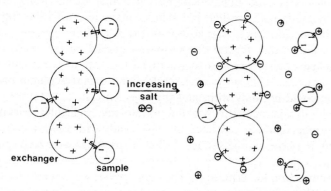

Fig. 4.1. A simplified representation of the dynamic equilibrium existing between an anion-exchange stationary phase and the ions in the mobile phase.

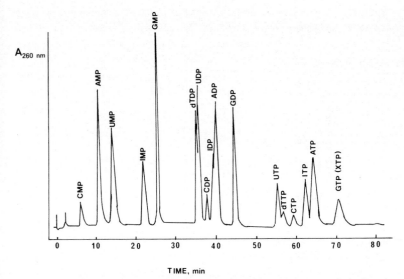

Fig. 4.2. Separation of mono- (MP), di (DP) and triphosphate (TP) nucleotides of adenine (A), guanine (G), hypoxanthine (I), xanthine (X), cytosine (C), uracil (U) and thymine (T). Chromatographic conditions: column, Partisil 10-SAX; mobile phase, linear gradient of 0.007 M KH_2PO_4, pH 4.0 to 0.25 M KH_2PO_4, 0.5 M KCl, pH 4.5 in 45 min; flow rate, 1.5 ml/min; temperature, ambient; detection, UV at 260 nm. Reproduced from Zakaria and Brown (1981), with permission.

retention from the weakly charged nucleotide monophosphates to the more highly charged nucleotide triphosphates.

The interaction of solute molecules with the ion-exchange stationary phase can be regarded as a sequential two-step process. Initially the solute must diffuse from the mobile phase (usually aqueous) into the stationary phase (often organic). The distribution between the two phases is largely responsible for the retention of a particular solute. Secondly, the solute must interact with, and diffuse through, the stationary phase.

On ion-exchange resins the rate of diffusion through the stationary phase, to and from the ion-exchange site, is relatively slow. This slow mass transfer results in a large contribution towards band broadening. Pellicular materials with only a thin layer of resin were evolved to

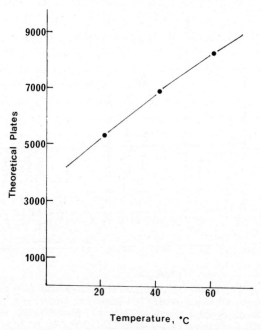

Fig. 4.3. Effect of temperature on column efficiency for the resolution of fumaric acid. Chromatographic conditions: column, Aminex HPX-87 (300×7.8 mm I.D.); Mobile phase, 0.026 M H_2SO_4; flow rate, 0.6 ml/min. Reproduced from Wood et al. (1980), with permission.

overcome this problem such that ions only needed to travel a relatively small distance. Unfortunately, the relatively large, inert glass spheres possess a low ion-exchange capacity and therefore silica-based ion-exchangers are much better, but control of mobile phase velocity and temperature remain important. Separations are often carried out at up to 60°C where the reduction in mobile phase viscosity allows more efficient separations (Fig. 4.3).

Most bonded phase ion-exchangers are stable up to 60°C and the speed of modern HPLC minimises the possibility of solute degradation. The k' values for solutes frequently respond differently to changes in temperature and consequently this parameter can be used very effectively to modify the selectivity of the column.

4.3. Effects of pH

4.3.1. Mobile phase effects

Many biochemical compounds which are charged at a specific pH can be separated by ion-exchange chromatography. In general, the acid and base equilibria which are established can be represented by the following equations where HA represents an acid and B represents a base:

$$HA \rightleftharpoons H^+ + A^- \qquad (4.1)$$

$$B + H^+ \rightleftharpoons BH^+ \qquad (4.2)$$

The degree of solute retention can be finely controlled by the pH of the mobile phase. At low pH, high concentrations of H^+ will shift the reaction equilibrium (4.1) to the left and only low concentrations of A^- will compete for anion-exchange sites. Consequently, a drop in pH will cause acidic solutes to elute rapidly from an anion-exchange column. Similarly at high pH, low concentrations of H^+ will shift the reaction equilibrium (4.2) to the left and only low concentrations of BH^+ will compete for cation-exchange sites. Thus raising the pH will cause basic solutes to elute rapidly from a cation-exchange column. This is shown in Fig. 4.4 where the effect of pH on retention time for several basic amino acids has been demonstrated. Under extreme conditions of pH, when all the molecules are either totally charged or uncharged, pH will have no further effect. For an acidic compound therefore:

retention is approximately proportional to $1 / [H^+]$

and it can be shown that for an acid of $pK_a = 5$ the following relationship exists between pH and relative retention by anion-exchange:

pH	retention
3	0.01
4	0.09
5	0.5
6	0.91
7	0.99

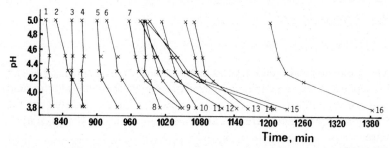

Fig. 4.4. Resolution of basic amino acids as a function of pH. Chromatographic conditions: column, Dowex 50, 17.5 μm (125×0.636 cm); mobile phase, 0.267 M citrate ion, 0.8 M sodium ion; flow rate, 0.5 ml/min; temperature, 60 °C; detection, ninhydrin derivatives at 540 nm and 440 nm. 1, β-aminoisobutyric acid; 2, γ-aminobutyric acid; 3, ethanolamine; 4, ammonia; 5, hydroxylysine; 6, ornithine; 7, lysine; 8, 1-methyl-histidine; 9, histidine; 10, 3-methylhistidine; 11, anserine; 12, 2-amino-3-guanidinopropionic acid; 13, canavanine; 14, tryptophan; 15, carnosine; 16, arginine. Reproduced from Hamilton (1963), with permission.

It should be noted that when pH is equal to the pK_a value, exactly half of the molecules are in the dissociated form, i.e. from equation (4.1) the concentration of HA, [HA], equals the concentration of A⁻, [A⁻]. It can be shown that:

$$pH = pK_a + \log_{10}([A^-]/[HA]) \qquad (4.3)$$

An inverse relationship can be demonstrated for bases on a cation-exchange column.

These relationships show that pH can be used with great effect to change the selectivity of the chromatographic system for various solutes. A pH value between the pK_a's of two solutes will give rise to the greatest selectivity. When choosing the pH value of the mobile phase it is useful to have some idea of the pK_a values of the components within the sample and a rough idea can be obtained from a consideration of the functional groups. Some of the more common functional group pK_a values are given in Table 4.1. A much more extensive listing can be found in the literature (CRC Handbook) and a guide to their prediction has been published (Perrin et al., 1980).

Compounds which carry both positively and negatively charged

TABLE 4.1
pK_a values of some common functional groups

Group	Acid pK_a	Basic pK_a
Aromatic amine		4– 7
Aliphatic amine		9–11
Carboxylic acid	4– 5	
Amino acid	2– 4	9–12
Phenols	10–12	

groups are termed amphoteric and their net charge is dependent upon the pH. At a certain pH the compound will possess an overall charge of zero and this pH is termed the isoelectric point (pI). At pH values below the pI value the compound will bind to cation-exchange stationary phases and at pH values above the pI value it will bind to an anion-exchange stationary phase. These considerations are particularly relevant to proteins.

Some useful buffers for controlling mobile phase pH value in cation-exchange chromatography are given in Table 4.2. Some useful

TABLE 4.2
Examples of cation-exchange buffers

Buffer	pK_a	pH range
Phosphate	2.1	1.1– 3.1
	7.2	6.2– 8.2
	12.3	11.8–12.0
Malonate	2.9	2.4– 3.4
	5.7	5.0– 6.0
Citrate	3.1	2.1– 4.1
	4.7	3.7– 5.7
	5.4	4.4– 6.4
Formate	3.8	2.8– 4.8
Acetate	4.8	3.8– 5.8
MES	6.2	5.5– 6.7
HEPES	7.6	7.6– 8.2
BICINE	8.4	8.2– 8.7
Borate	9.2	8.2–10.2

TABLE 4.3
Examples of anion-exchange buffers

Buffer	pK_a	pH range
Piperazine	5.7	5.0– 6.0
bis-Tris	6.5	5.8– 6.4
bis-Tris propane	6.8	6.4– 7.3
Triethanolamine	7.8	7.3– 7.7
Tris	8.1	7.6– 8.0
Diethanolamine	8.9	8.4– 8.8
Ethanolamine	9.5	9.0– 9.5
Piperazine	9.7	9.5– 9.8
Piperidine	11.1	10.6–11.6
Phosphate	12.3	11.8–12.0

buffers for controlling mobile phase pH value in anion-exchange chromatography are given in Table 4.3.

4.3.2. Stationary phase effects

The degree to which the ion-exchange stationary phase can interact with the solute molecules is affected by pH. At pH values above the pK_a of the ion-exchange groups, anion-exchangers are neutralised by base:

$$R^+ + OH^- \rightarrow ROH \qquad (4.4)$$

whereas at pH values below the pK_a of cation-exchangers the ion-exchange groups are protonated (neutralised):

$$R^- + H^+ \rightarrow RH \qquad (4.5)$$

Thus variation of pH will affect the ion-exchange capacity of the stationary phase. This variation in ion-exchange stationary phases must be used at the plateau portion of their pH curves (Fig. 4.5). Both strong anion- and strong cation-exchangers can be used over a wider pH range than the weak ion-exchangers and consequently they have found a wider applicability in difficult separations. The weak ion-exchangers are of more value for highly charged molecules which are otherwise too strongly retained.

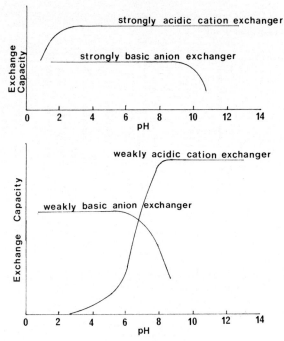

Fig. 4.5. Effect of pH on the capacity of ion-exchange supports. Reproduced from Scott (1971), with permission.

The interaction of solutes with the ion-exchange sites on the stationary phase is not the only factor which can influence the degree of retention. For example, the overall organic nature of the surface of coated silicas or the organic bulk of polystyrene–divinylbenzene supports can influence retention through a partitioning effect with non-polar molecules. Similarly, underivatised silanol groups on modified silica supports can affect retention by adsorption effects with polar solutes (Asmus et al., 1976). These effects give rise to separations in which competing mechanisms make detailed interpretation of the specific factors affecting separations extremely complex. Solutes which are not amenable to simple ion-exchange chromatography can be

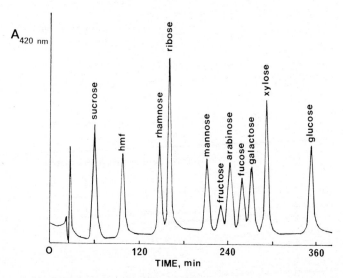

Fig. 4.6. Anion-exchange chromatography of carbohydrates as their borate complexes. Chromatographic conditions: column, Aminex A-28 10 μm (50×0.3 cm I.D.); mobile phase, gradient of 40 ml of 0.124 M borate, pH 7.0 to 40 ml of 0.58 M borate, pH 10.0; flow rate, 10 ml/h; temperature, 55°C; detection, orcinol reagent reacted at 100°C for 20 min and measured by UV at 420 nm. Reproduced from Morrison et al. (1976), with permission.

separated through formation of ionic complexes. For example, sugars can be separated as their borate anion complexes (Fig. 4.6).

4.4. Stationary phases

A wide variety of ion-exchange packings are commercially available from such companies as Pharmacia, SynChropak, Toyo Soda, Bio-Rad, Whatman and Varian. They can generally be divided into 4 classes:
(1) Cross-linked polystyrene-based, 5–25 μm.
(2) Pellicular, 30–50 μm.
(3) Silica-based, 5–10 μm.
(4) Hydrophilic gel, 10 μm.

4.4.1. Cross-linked polystyrene stationary phases

The percentage of cross-linking in the polystyrene resins is typically in the range of 1–16% but averages approximately 8%. Cross-linking improves the mechanical strength, reduces swelling and improves selectivity. However, such supports exhibit a low rate of mass transfer and elevated temperatures (60°C) are often employed to improve column efficiency. Cross-linking also reduces the pore size which is typically in the range of 1–5 nm. This means that most large molecules are excluded from the resin and limits the use of such supports. It follows that all resin-based ion-exchangers also separate by size, regardless of whether the solute molecules are charged. Typical values for the capacity of such supports lie in the range 10^3–10^4 μEq/g.

4.4.2. Pellicular stationary phases

Pellicular packings generally utilise a glass bead support and are usually of the size 30–50 μm. These supports suffer from a very low capacity (in the range of 5–50 μEq/g). They also give low k' values and require smaller sample sizes.

4.4.3. Silica-based stationary phases

In theory, silica-based bonded phases have only a monolayer of functional groups bound to their surface. The capacity of these supports is quite good and is usually in the range 10^2–10^3 μEq/g. The rapid diffusion which is obtainable with such a thin surface layer results in considerable improvements in column efficiency compared to totally polymeric supports. Typical particle sizes are in the range 5–10 μm. Recent developments in silica-based ion-exchangers have led to materials with a large pore diameter and these have found particular application in separations involving large biological molecules such as proteins.

4.4.4. Hydrophilic gels

These supports have recently been introduced to the market and comprise a hydrophilic polyether-based material of large pore size, to which either diethylaminoethyl (anion-exchange) or sulphopropyl (cation-exchange) groups have been attached (Kato et al., 1983b, 1984). The porous nature of these supports means that high loading capacities are possible. They have found wide application for polypeptide separations using water–organic solvent–buffer mobile phases.

The choice of ion-exchange packing depends on the parameters described above, but for some compounds such as amino acids and proteins, considerable advantages are gained by exploiting their amphoteric nature. Large changes in selectivity will be observed between a separation on anion-exchange resin to that on a cation-exchange resin.

4.5. Mobile phases

In ion-exchange chromatography the mobile phase is usually an aqueous solution of buffered salts to which an additional miscible organic liquid such as methanol, acetonitrile or dioxane is often added. Where an aqueous organic mixture is being used the composition of the mobile phase within the pores of the stationary phase is often biased towards a lower concentration of organic phase and arises through the affinity of the ion-exchange sites for water. Alternatively, in the presence of salts it is possible that the organic solvent will be 'salted out' into the stationary phase giving rise to the reverse effect. These variations in mobile phase composition can be utilised for the separation of non-ionic solutes such as sugars (Paart and Samuelson, 1973) and aromatic hydrocarbons (Ordermann and Walton, 1976; Dieter and Walton, 1983).

In ion-exchange chromatography the ionic strength of the mobile phase can be used to manipulate the retention of a particular solute by competing for the ion-exchange sites. Thus the solvent strength increases as the ionic strength increases and it can be shown that:

k' **is approximately proportional to** $1 /$ **ionic strength**

This relationship holds true in both cation and anion-exchange chromatography, but only when the ions are monovalent. For divalent sample ions:

k' **is approximately proportional to** $1 / ($ **ionic strength** $)^2$

A change in ionic strength affects the retention of all ions of the same charge equally but has no influence on the solvent selectivity. Consequently, this method can be used to influence the solvent selectivity for sample ions of different charge. However, it should be noted that in the case of 'salting out' it is possible for an increase in ionic strength to cause an increase in retention.

Solvent strength in ion-exchange chromatography is also dependent upon the type of ions in the mobile phase. The list below represents a typical example of those ions which, in equal concentration, cause an increase in solvent strength:

(a) For anion-exchange:

$$F^- < OH^- < acetate^- < formate^- < Cl^- < Br^- < I^- < NO_3^-$$

$$< oxalate^- < SO_4^{2-} < citrate^-$$

(b) For cation-exchange:

$$Li^+ < H^+ < Na^+ < NH_4^+ < K^+ < Rb^+ < Cs^+ < Ag^+ < Tl^+ < Mg^{2+}$$

$$< Zn^{2+} < Co^{2+} < Cu^{2+} < Cd^{2+} < Ni^{2+} < Ca^{2+} < Sr^{2+} < Pb^{2+} < Ba^{2+}$$

In practice the use of halide ions in conjunction with stainless steel columns should be strictly limited because of their corrosive action. This is particularly marked when halide ion solutions are left for extended periods in contact with stainless steel. In this context it is worth noting the introduction of ceramic pistons in some newer HPLC systems to allow the use of halide ions in separations.

In addition to its effect on solvent strength, a change of mobile phase ion can sometimes be used to alter the solvent selectivity.

4.6. Added solvent

The addition of an organic solvent to modify the mobile phase is often used to alter the selectivity of the stationary phase. Methanol, acetonitrile, ethanol and dioxane at concentrations of up to 10% are used for this purpose. The effects resulting from such an addition are similar to those observed in reversed phase HPLC and generally result in a reduction in the retention of the solutes.

CHAPTER 5

High performance size exclusion chromatography

5.1. Introduction

One of the earliest concepts to be realised by researchers was the possibility of separating molecules from complex mixtures on the basis of their molecular size. Thus the techniques of ultrafiltration and dialysis have been developed to achieve such separations. However, higher resolution and selectivity can be obtained using size exclusion chromatography (SEC). The development of suitable chromatographic supports was addressed by biochemists and organic chemists independently, who developed two separate types of stationary phase and also two sets of nomenclature. Gel filtration, as defined by the biochemist, refers to macroporous cross-linked dextrans and acrylamides compatible with aqueous buffers (Porath and Flodin, 1959). Gel permeation, as defined by the organic chemist, refers to the use of rigid, polymer-based resins such as polymethacrylate, polydivinyl benzene and microparticulate silica resins which are stable in both organic and aqueous buffers at pH values less than 8.0. The term size exclusion is used in this text to encompass the forms of chromatography otherwise known as gel filtration and gel permeation. The development of high performance (implying faster analysis, better resolution) size exclusion supports, particularly for proteins, has been relatively slow, mainly because of the difficulties inherent in the principle upon which the separation is based.

5.2. Principles

If a mixture of molecules with different sizes is applied to a column which has been packed with porous material of a specific pore size, only a certain percentage of the molecules will be small enough to enter the pores, the rest being excluded. The larger molecules will therefore spend less time in the pores and will pass through the column more rapidly. The separation is based on the physical parameters of the molecules themselves and not on their chemical properties and it is this principle which differentiates SEC from the other forms of chromatography. Separation of two compounds occurs because of the differences in the distribution of the molecules between the mobile phase and the stationary phase. Mathematical treatments of the theory of size exclusion chromatography are usually based upon models which consider the solute molecules as spherical and the stationary phase as a rigid gel (Ambler and MacIntyre, 1975), although in reality not all the molecules are the same shape in solution (Table 5.1). It has been demonstrated that a polymer of a given molecular weight can have a radius equivalent to that of a globular protein of much higher molecular weight (Ui, 1979).

A theoretical analysis of SEC can be derived from the chromatographic elution profile. The retention of a given molecule is expressed as the distribution coefficient K_d (Fig. 5.1), where V_o is the volume of the solvent outside the particles of the stationary phase and V_i is the solvent inside the particles which is available for chromatography.

TABLE 5.1
Illustration of the mass (kDa) and Stokes radii (Å) of ferritin, catalase and bovine albumin, and the mass of a polymer of an equivalent radius

	Mass (kDa)	Radius (Å)	Equivalent polymer mass (kDa)
Ferritin	480	156	57.3
Catalase	240	104	28.7
Bovine albumin	67	70	14.3

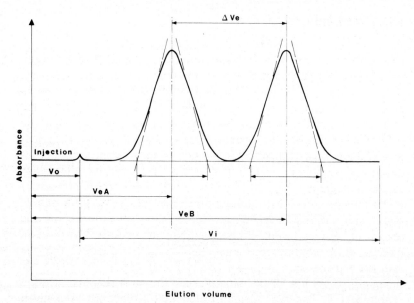

Fig. 5.1. Diagrammatic illustration of the parameters involved in size exclusion chromatography. V_o, volume of solvent outside the particles; V_i, volume of solvent inside the particles available for chromatography, V_e (A), elution volume of a sample A; V_e (B), elution volume of a sample B. From this, the retention coefficient K_d can be derived for each sample:

$$K_d(A) = \frac{V_e(A) - V_o}{V_i}$$

Reproduced from LKB technical brochure, with permission.

allows the determination of the elution volume (V_e) of a given molecule of interest. If the solute is large and excluded from the gel then $V_e = V_o$ and $K_d = 0$. If the solute is small enough to penetrate completely into all the accessible pore volume, then $V_e = V_o + V_i$ and $K_d = 1$. In practice the value of K_d varies between 0 and 1 since not all the pore volume is ever completely accessible. For any given column there will be a fixed relationship between these parameters and the elution volume; however, the value of K_d is independent of the column parameters (i.e. height, diameter).

5.3. Factors affecting resolution

5.3.1. Selectivity

Since the principle of SEC is different from the other forms of chromatography the parameters of the resolution equation are not normally applicable. However the control of selectivity in SEC is simple and predictable once the parameters of the stationary phase are optimised. The critical features of the stationary phase are pore volume, pore size distribution and particle shape. It is also important that the support is inert to both ionic and hydrophobic interactions with the sample molecules and mobile phase. Developments to optimise particle shape and size distribution have led to the increasingly popular use of small particles (5–100 μm), spherical in shape and with a minimum size distribution of plus or minus 10% (Chapter 10). Strict control of these criteria is essential for the manipulation of the physical aspects of column efficiency. An excess in pore size distribution results in increased peak broadening (Werner and Halasz, 1980). Soft dextrans and agaroses which lack mechanical stability are not suitable for use in HPLC because of the pressures applied. A list of commercially available supports currently being used can be found in the literature (Lesec, 1985; Wehr, 1984). The two most common stationary phases for aqueous SEC are cross-linked polymer-based polyether or polyester and silica-based phases. Each of these stationary phases have hydroxyl groups covalently linked to the surface (Regnier and Noel, 1976). For non-aqueous SEC the silica-based phases are popular, e.g. ZORBAX SEC, a 'silanised' porous silica microsphere (Yau et al., 1978). Alternatively, polystyrene divinyl benzene stationary phases have been used and are now available in a wide range of efficiencies and pore sizes, e.g. TSK H (Cooper et al., 1975). At present the two most popular stationary phases are silica based to which an organic phase has been attached, i.e. glycerol-propyl type, Waters I series, Synchropak GPC, TSK SW types (Kato et al., 1980). These stationary phases possess properties suitable for use with aqueous buffers. The most popular stationary phases for use with non-aqueous buffers are of the type PSDV (Spheron, Shodex) and methacrylate (Spheron P series), which may also be used with aqueous solvents. TSK PW polyether supports are very popular for separation of proteins and polynucleotides. Pore sizes of 25–4000 Å are available. Pore sizes larger than this tend to show reduced resolution due to decreased plate number.

5.3.2. Column capacity

The disadvantage of SEC compared to the other forms of chromatography is the relatively low sample capacity. In analytical separations where optimum resolution is required, the total load volume should not exceed 1–2% of the total column volume (Roumeliotis and Unger, 1979). This need not apply in all circumstances since when using SEC for the purpose of changing buffers or desalting a sample up to 10% of the column volume is easily accommodated.

5.3.3. Pore size distribution

A critical feature affecting selectivity in SEC is the minimisation of pore size distribution. Control of this parameter facilitates separation of molecules with a particular size distribution. Stationary phases are available over a wide range of pore sizes (Table 5.2) and often these columns can be used in series (Mori, 1979). Thus, effective selection within a broad range is accomplished by the first column and fractionation within a more defined range is achieved on the second column.

TABLE 5.2
The molecular weight range (in Daltons) for the separation of proteins and 'random coil' molecules using currently available supports

Support	Globular proteins	Random coil
TSK 2000	1,000– 30,000	500– 8,000
TSK 3000	2,000– 80,000	1,000– 30,000
TSK 4000	20,000–1,000,000	20,000–150,000

5.3.4. Pore volume

Pore volume (V_i) is an important parameter affecting resolution in SEC. This parameter presents difficulties in the design of stationary phases since a large pore volume is usually achieved using an open lattice bead structure which will be less mechanically stable to increased pressures at high flow rates. The compromise between pore volume and mechanical stability has been the major difficulty in the development of supports suitable for macromolecules. Thus, analytical size exclusion columns generally are longer than those used in the other chromatographic modes, so that the amount of stationary phase and thus the effective pore volume available for chromatography is increased.

The migration of molecules between the stationary phase and the mobile phase is driven by random movement or diffusion, a factor which is deleterious to high resolution in all forms of chromatography. Since resolution in SEC is solely dependent on diffusion, unlike the other forms of chromatography, optimisation of stationary phase particles is important to improve mass transfer. Factors deleterious to mass transfer can be divided into three separate types: those attributable to stagnant mobile phase in the pores of the particles, those caused by differential penetration of the solute molecules into the stationary phase and, finally, longitudinal diffusion between the particles (Snyder and Kirkland, 1979).

Even within the best supports available there will be some irregularities in the particle shapes which will lead to non-uniform channels through which the molecules will permeate. Optimisation of the physical parameters of the column, the type of stationary phase and the method of packing are normally directed at overcoming these effects.

5.4. Mobile phase effects

5.4.1. pH

The mobile phase can affect resolution through direct interaction with both the stationary phase and the solute. Silica columns are best used at pH values less than 7.0 to minimise the formation of silanolate anions which could then participate as weak cation exchangers. Silanisation of silica-based stationary phases is essential to prevent any interactions of either aqueous mobile phases or sample molecules with free silanolate anions.

5.4.2. Ionic strength

SEC of macromolecules is generally carried out using buffers which contain counter-ions to stabilise charged residues and thus maintain the structural integrity of the solute. This is particularly important for proteins, which contain a wide variety of interacting groups. However, an excessively high ionic strength will tend to promote hydrophobic interactions and therefore a low ionic strength buffer (0.1 M) is normally recommended. The inclusion of suitable counter-ions (Na^+, K^+, NH_4^+), will also offset some of the problems associated with high pH values by masking reactive silanolate anions.

By altering the pH in conditions of low ionic strength, the polarity of the stationary phase can be altered to suit the particular requirements of the solute. This has been called 'non-ideal' SEC and is useful for certain proteins (Kopaciewicz and Regnier, 1982).

5.4.3. Flow rate

Mobile phase flow rates of around 0.5–1.0 ml/min are recommended for the resolution of a variety of macromolecules on a number of stationary phases. However, flow rates of 0.1–0.5 ml/min are recommended for use with TSK SW and H types. In general, for larger molecules (polynucleotides, proteins), the mass transfer term is much larger and the flow rate has to be correspondingly reduced to maintain resolution.

5.5. Molecular weight determination

The size of the solute molecules is characterised by their hydrodynamic radius (Stokes radius) in a particular solvent. Using SEC it is

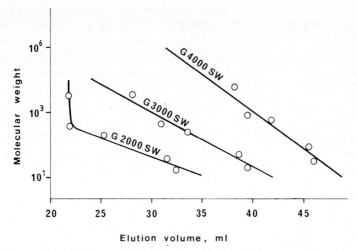

Fig. 5.2. A plot of the retention volume versus the logarithm of the molecular weight of a series of standards (proteins) of known molecular weight. Chromatographic conditions: columns, TSK-Gel G2000 SW, G3000 SW and G4000 SW (600×7.5 mm I.D.); mobile phase, 1/15 phosphate buffer, 1/10 M KCl; flow rate, 1 ml/min; temperature, ambient. Reproduced from TSK technical bulletin, with permission.

possible to estimate the molecular weight of an unknown protein or polymer by comparing its retention volume with a plot of retention volume versus the logarithm of molecular weight for a series of standards of known molecular weight (Fig. 5.2). These plots can provide accurate determinations of molecular weights for polymers and proteins provided the unknowns adopt a similar conformation in solution to that of the standard. This implies an inherent error and calculations based solely on this type of analysis require further confirmation.

For a more accurate assessment, strong denaturants such as 0.1% sodium dodecylsulphate (SDS) or 6 M guanidine hydrochloride can be included in the mobile phase in aqueous SEC. This tends to promote uniform conformation of the components of the solute and will be closer to the ideal situation. The inclusion of denaturants has the added benefit of reducing any chemical interactions of solutes the stationary phase. The addition of detergent to the mobile phase promotes random coil conformation and therefore a corresponding increase in the Stokes radius occurs. This has been useful for the estimation of molecular weights of both fibrous and globular proteins (Barden, 1983), although some subunits of multimeric enzymes will dissociate under these conditions. A comparison of SEC in SDS and SDS–polyacrylamide gels for the determination of molecular weight concluded that the former is more useful when a detailed knowledge of the solute is available (Josic et al., 1984). Alternatively, charge interactions can be neutralised completely by chemical modification, and by also disrupting hydrophobic interactions, the solute should behave ideally (Meredith, 1984).

Early supports for high performance SEC proved disappointing in terms of stability, resolution and recovery. Recent advances have offset many of these problems and offer researchers another powerful fractionation tool. There are distinct advantages over the other forms of chromatography since recovery is generally in excess of 90%, the mobile phase is simple (requiring no complex buffer change or gradient), the profile is highly reproducible and the elution order is predictable.

CHAPTER 6

High performance normal phase chromatography

6.1. Introduction

The term normal phase chromatography describes one of two sub-classes of liquid–solid phase chromatography, these being normal phase (NPC) and reversed phase (RPC) chromatography. Normal phase chromatography is carried out on stationary phases which in physicochemical terms are relatively polar, in contrast to reversed phase supports which are essentially non-polar. Thus NPC can be considered as the chromatography of a solute on a polar stationary phase whose retention is governed by polar interactions between the stationary phase and mobile phase. A generalised theory of retention is difficult to present, except in extreme situations, because of the complex interactions which can occur between the solute and solvent molecules. Contributions to these interactions from dispersion forces, dipole moments, hydrogen bonding and dielectric interactions between ions may also occur (Chapter 2). At present, two significant theoretical models have been proposed to explain the retention mechanism in NPC; these are known as the displacement model (Snyder, 1968; Soczewinski, 1968) and the sorption model (Scott and Kucera, 1973). For a detailed comparison of the two models, the reader is referred elsewhere (Snyder, 1974, 1983; Snyder and Poppe, 1980). In this chapter only a brief overview of each of the models will be discussed.

6.2. Displacement model

The majority of experimental investigations into the retention mechanism of NPC have been carried out using silica as the stationary phase although more recently these studies have been extended to include other polar bonded supports (amino phase) (Snyder and Schunk, 1982). Experimental observations of retention using mobile phases of differing polarity have resulted in the proposal of a model which suggests that a monolayer of mobile phase molecules is adsorbed to the surface of the stationary phase. The eluting solute binds to the stationary phase and thereby displaces (or competes off) one or more solvents molecules (Fig. 6.1). The retention sites on the stationary phase are silanol (SiOH) groups for silica, AlOH for alumina, and amino (NH_2) groups for the amino bonded phase. A thermodynamic interpretation concludes that the interaction of the solute molecules with the stationary phase is the dominant interaction and that the interaction of the solute with the mobile phase is relatively unimportant.

A relationship between solvent strength and retention time can be derived as follows:

$$\log k'_1 / \log k'_2 = \alpha' \times A_s \times (E_1 - E_2) \qquad (6.1)$$

where k'_1 and k'_2 are the capacity factors of a given solute in two different mobile phases, α' is a function of the strength of the adsorbant surface, and E_1 and E_2 are the values of the solvent

Fig. 6.1. Schematic illustration of the displacement model. Retention is determined by the displacement of a mobile phase molecule (M), by a solute molecule (S), from a binding site on the stationary phase (⊥). Adapted from Snyder and Poppe (1980), with permission.

strength of the different mobile phases (Snyder, 1983). Thus there is a linear relationship between retention and solvent strength. The slope of the plot of retention against solvent strength is proportional to A_s, the molecular mass of the solute. This equation has been verified for both non-polar solvents and solutes and it has been suggested that the deviations that occur within such plots are due to the various interactions between more polar groups: (a) the polar groups of the solvent which do not distribute themselves evenly over the surface (restricted access delocalisation); (b) lateral interaction of solvent molecules (site competition delocalisation); and (c) interaction of solute and solvent in the mobile phase (e.g. hydrogen bonding). This model has also been used to explain retention on both alumina and amino stationary phases (Snyder and Schunk, 1982). Cyanopropyl stationary phases show properties similar to those of ordinary silica when using non-polar mobile phases (possibly due to the presence of free silanol groups). In the presence of more polar mobile phases, it is thought that the cyano groups are the dominant reactive group (Weiser et al., 1984). A mathematical interpretation concludes that the displacement model can be used successfully to predict retention times for both silica and alumina but not so accurately for chemically bonded phases (De Ligny et al., 1984).

6.3. Sorption model

The sorption model was derived from studies to determine the quantity of solvent adsorbed to the surface of silica. Experimental evidence suggests that silica adsorbs water to its surface, some of which can be removed by either heating to $110\,^\circ$C, or by sequential organic phase extraction. More can be removed at very high temperatures, but this also results in the breakdown of silanol groups (Scott, 1982). Measurement of the adsorption of different concentrations of polar modifiers in an inert solvent allows the adsorption isotherms to be calculated (Fig. 6.2). When the concentration of the modifier is low, the isotherm fits closest to a monolayer function. When the polar modifier is present at a high concentration a bilayer adsorption isotherm function is produced:

$$y = A - (A + ABx/2)/(I + Bx + Cx^2) \qquad (6.2)$$

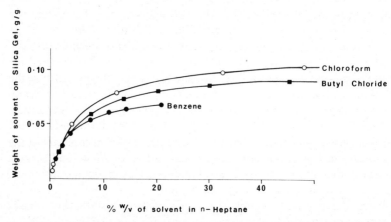

Fig. 6.2. A plot of the Langmuir adsorption isotherm for three non-polar solvents (chloroform, butyl chloride and benzene) in n-heptane. It can be seen that the data fit very closely to the monolayer function $y = x/(A + Bx)$. Reproduced from Scott and Kucera (1979).

It is proposed that a monolayer of solvent is adsorbed to the surface of the silica and that the solute is adsorbed onto the surface when the polarity is less than that of the solvent. No displacement or 'site competition' is envisaged. When the polar modifier concentration is high, a bilayer is formed and the solute then competes with a molecule of the polar solvent in the second layer (Fig. 6.3). Therefore, the major difference between the two models is the level of impor-

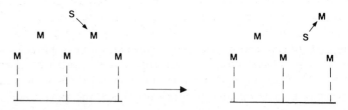

Fig. 6.3. Schematic illustration of the sorption model. Retention is determined by the replacement of a mobile phase molecule (M), by a solute molecule (S). The binding sites on the stationary phase ($\perp$) are blocked with a layer of mobile phase molecules. Adapted from Snyder and Poppe (1980), with permission.

tance placed on the interactions between the solute and the solvent. A detailed comparison of the two models has been made (Snyder, 1980).

Both of these models attempt to explain the relationship between solute retention and the polarity of the mobile phase. An independent study of mixed solvent theory has supported the sorption model (McCann et al., 1982) whereas another study on the effects of solvent composition on the chromatography of alkylphenols and naphthols shows evidence in favour of the displacement model (Hurtubise et al., 1981). At present, more evidence has been presented in support of the displacement model, although both have been used successfully to predict the effects of various mobile phase parameters.

6.4. Stationary phase effects

The use of small spherical particles for NPC results in plate numbers in excess of 10,000 which compares favourably with the efficiencies of stationary phases currently in use for reversed phase HPLC. A list of the common functional groups used in NPC are shown in Table 6.1. Most of the popular supports used in NPC are silica based since alumina undergoes undesirable side reactions (e.g. irreversible reaction with carboxylic acid). The introduction of bonded phase supports has led to the availability of many other polar stationary phases, some of which can also be used in the reversed phase mode (Lochmuller et al., 1979). A bonded diamino phase was found to be optimal for aromatic hydrocarbons (Chmielowiec and George, 1980).

TABLE 6.1

A list of the common functional groups used in NPC stationary phases

Non-bonded	Bonded
Silica	Amino (NH_2)
Alumina	Diamino ($NH(CH_2)2NH_2$)
	Methylamino ($NH(CH_3)_2$)
	Cyano (CN)
	Diol ($CH_2OH)_2$)
	Ether (CH_2OCH_3)

6.5. Mobile phase effects

In NPC, as with the other modes of chromatography, the mobile phase is used to control retention and selectivity. A list of the more commonly used solvents, in order of polarity, is given in Table 6.2. The most popular mobile phases are based on either hexane or heptane to which varying concentrations of a more polar solvent are added, e.g. propanol or ethyl acetate. In NPC the strength of the mobile phase increases with increasing polarity, resulting in a corresponding reduction in k'. (This is the opposite of that observed in RPC, where a decrease in polarity results in a decrease in k'.) In support of the displacement theory, a linear relationship between log k' and solvent strength (E_0) has been demonstrated using an amino bonded column (Snyder and Schunk, 1982).

Selectivity effects are determined by mobile phase interactions between the stationary phase and the solute. Those mobile phases containing proton donors will interact with basic solutes. Conversely, mobile phases containing proton donor acceptors will interact strongly with acidic solvents. Where excessive interactions occur between the solute and the mobile phase, peak tailing can result. However, the inclusion of a basic modifier (triethylamine) can overcome such strong interactions and thereby improve peak shape.

6.6. Temperature effects

Early observations on the effect of temperature in NPC indicated that, as a general rule, an increase in temperature resulted in a

TABLE 6.2

A list of the solvents used in NPC in order of increasing polarity
(reproduced from Snyder and Kirkland (1979), with permission)

Methanol	Ethyl acetate	1-Chlorobutane
Acetonitrile	Carbon tetrachloride	Hexane
Ethanol	Tetrahydrofuran	Heptane
Propanol	Dichloromethane	

decrease in retention; however, this generalisation is dependent on the specific polar modifier in the solvent; for example ethanol resulted in an increase in retention. More recent studies have indicated that the temperature effects are dependent on the nature of the stationary phase and that the most polar phase examined produced the largest decrease in retention with increase in temperature. An alternative explanation for the variable effect of temperature on retention has been attributed to the relative amounts of water adsorbed to the stationary phase, which in turn is related to the polarity of the

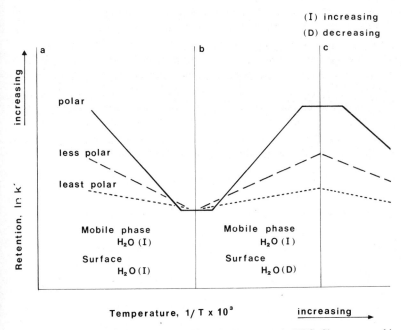

Fig. 6.4. Diagramatic illustration of the effect of temperature in NPC. Chromatographic conditions: stationary phases, polar (silica), less polar (cyanopropyl), least polar (n-butyl) (250×2.4 mm I.D.); mobile phase, n-hexane with the addition of water; flow rate, 1.6 ml/min; solutes, aniline, p-chloroaniline, 2,6-dimethylaniline, m-chloroaniline, n-methylaniline, 2,6-dimethylphenol, N,N-dimethylaniline, o-chloroaniline and benzene. The figure is divided into three different temperature areas (a,b,c). Reproduced from Sisco and Gilpin (1980).

stationary phase. A summary of these effects is shown in Fig. 6.4 (Sisco and Gilpin, 1980). However, in general, temperatures above 40°C result in an increase in retention, whereas below this, there is a decrease in retention.

6.7. Other considerations

Normal phase stationary phases are generally less stable than reversed phase stationary phases. A common finding is the deterioration of amino columns through the formation of Schiff bases with carbonyl groups of the solute, although a method of reactivation has been suggested (Karlesky et al., 1981). The main advantage of NPC is the ability to separate those species that are only sparingly soluble in water, and also those species that are very hydrophilic and which are not retained on a reversed phase support (Abbott, 1980). Furthermore, the availability of NPC supports allows the hydrophilicity of the sample to be exploited as another feature of selectivity. Although NPC is not as popular as RPC, it is being used increasingly with the realisation that many biomolecules, hitherto regarded as unstable in high concentrations of organic solvent, can be efficiently chromatographed using NPC (e.g. human interferon, Chapter 11).

CHAPTER 7

High performance reversed phase chromatography

7.1. Introduction

Of all the chromatographic techniques used, reversed phase HPLC is by far the most popular with at least 60% of all analytical separations carried out in this chromatographic mode. The term reversed phase chromatography was originally coined by Howard and Martin (1950) who carried out liquid–liquid chromatography on a stationary phase of paraffin oil and n-octane with aqueous eluents. In such a partition system the conventional methodology, which used a polar stationary phase and a less polar mobile phase, was reversed, with the mobile phase being more polar than the stationary phase.

Reversed phase chromatography differs from most other chromatographic techniques in which the attractive forces between the stationary phase and the mobile phase are dominant. In reversed phase chromatography the stationary phase is generally an inert hydrocarbon and usually the only interactions with the solute are hydrophobic, with selectivity being dominated by solvent effects. Typically, the mobile phases used in reversed phase chromatography consist of either water alone, or water plus an organic modifier, although more recently much interest has been expressed in non-aqueous mobile phases. These different mobile phases have resulted in the acronyms PARP, MARP and NARP which stand for plain aqueous, mixed aqueous and non-aqueous reversed phase chromatography, respectively.

While variations to the general mode of reversed phase chromatography are abundant and rapid developments continue in the area, a general outline of the technique may be given which will be expanded in later sections in this chapter. The stationary phase is usually composed of spherical silica particles to which *n*-octadecyl ligands have been covalently bonded. Other popular ligands are *n*-octyl and phenyl. The silica microparticles usually have a diameter of less than 25 μm for analytical applications, although in preparative HPLC this diameter is often larger. The stationary phase is usually packed into stainless steel analytical columns which have an internal diameter of 4–5 mm and are 50–300 mm long, but there is a recent trend towards shorter columns and smaller diameter packings. The most popular mobile phases are either aqueous or aqueous plus an organic modifier such as methanol or acetonitrile. The mobile phases are usually pumped through an analytical column at 0.5–2 ml/min whilst flow rates through preparative columns are usually much higher.

The popularity of the reversed phase mode is understandable in view of the numerous advantages it has for the potential chromatographer. The greatest of these advantages is the inertness of the stationary phase, which allows the exploitation of a wide range of solvent effects through variation of the mobile phase composition; for example, the addition of salts and the alteration of mobile phase pH and temperature.

In the present chapter each of the aspects of reversed phase chromatography which have been discussed above will be considered in more detail with the emphasis on guiding the potential chromatographer through the many possible combinations of mobile phase and stationary phase.

7.2. Stationary phases

The most commonly used stationary phase in reversed phase chromatography consists of octadecylsilane groups (ODS) covalently bonded to silica; however, a number of different types of ODS packings of varying chromatographic character are utilised and these will be considered in this section. In addition, a number of different functional groups besides ODS can be attached to silica and some of the more popular of these will also be discussed.

7.2.1. Octadecylsilane (ODS) packings

ODS packings may be broadly divided into monomeric and polymeric phases. Monomeric phases generally have a more homogeneous coverage of the silica by the bonded ligand and have a relatively lower carbon loading which can result in a high proportion of free silanol groups. Thus, monomeric packings which are not fully end-capped can display a mixed mechanism of chromatographic separation due to the occurrence of partition effects in addition to conventional solvophobic interactions. Monomeric phases are generally more efficient than polymeric ODS phases for the separation of polar solutes.

Polymeric phases usually have a relatively high carbon loading but show poor mass transfer characteristics for polar compounds due to the nature of the dense hydrophobic chains. Polymeric phases are generally recommended for the separation of solutes of intermediate polarity and also for non-polar compounds. Furthermore, since polymeric phases have a higher percentage carbon loading, the columns have a high sample loading capacity and may be more useful for preparative chromatography.

7.2.2. C_2–C_8 packings

Packings with alkylsilane ligands, where the alkyl groups vary between 2 and 8 carbon atoms, are usually used after a chromatographic separation has been previously developed on an ODS support. The advantages conferred by short alkyl chains are that retention of compounds is generally reduced so that a comparable k' value to an ODS packing can be obtained by reduction in the content of the relatively expensive organic modifier in the mobile phase. Also, better peak symmetry has been reported using C_8 packings, which may be attributed to improved wetting of the surface of the stationary phase due to an improvement in solute mass transfer. In general, C_8 phases are the most popular alternative alkylsilane packing and the very short chain C_2 ligands are only occasionally used because of the comparative difficulty in optimising separations due to the mixed separation mechanisms resulting from the increase in solute–surface interactions at the silica support.

7.2.3. Phenyl packings

Packings with phenyl groups covalently bonded to the silica support usually show a reduced retention for aliphatic compounds compared with either C_8 or ODS phases under comparable chromatographic conditions and this is attributable to the reduced carbon loading of this phase. A relative increase in the selectivity of phenyl phases compared with ODS phases for aromatic solutes with respect to alkyl solutes of the same molecular weight and polarity has been reported and results from pi–pi interactions between the phenyl ligand and the aromatic solute (Tanaka et al., 1982). A further improvement in the selectivity of this type of packing over long chain alkyl packings has been reported for solutes which have a restricted planar conformation.

7.3. Factors affecting separations

Recent years have seen a rapid expansion in the literature concerning the separation mechanisms in reversed phase liquid chromatography and in the physical parameters which may modify chromatographic separations. In this section we shall first consider some of the general concepts underlying reversed phase chromatography and follow this with a consideration of the theories underlying separation mechanisms. The practical application of this theory to physical parameters including solvent composition, temperature, pH, buffer composition and sample loading will then be presented to allow practising chromatographers to optimise their chromatography conditions by modification of these parameters.

7.3.1. Basic concepts

The interactions between the stationary phase, the mobile phase and the solute in reversed phase liquid chromatography may be considered to be by either partition or adsorption, or a mixture of both. In adsorption the solute and the mobile phase compete for sites at an interfacial region on the stationary phase. In this sense reversed phase liquid chromatography becomes very much like liquid–liquid chromatography. In partition, the solute partitions between the mobile phase and the stationary phase with no consideration being given to the interfacial region; under these conditions the solute–stationary phase interactions are defined by weak, non-specific Van der Waals forces in which the mobile phase is dominant. Experiments to date suggest that reversed phase liquid chromatography is not a simple mechanism but probably a mixture of mechanisms, and this has resulted in a third mechanism being proposed (Knox and Pryde, 1975) whereby the organic modifier in the mobile phase is adsorbed onto the stationary phase, thus creating a new stationary phase. The solute molecules then partition between the mobile phase and the newly created stationary phase.

The major problem in defining the separation mechanism resides in a lack of understanding of the nature of the surface coverage of the bonded groups on the stationary phase. The simplest description of the surface is that it consists of a brush-like structure (Karch et al., 1976) with individual alkyl bristles of the bonded phase protruding into the mobile phase. However, it has been recognised for some time that alkyl chains can undergo hydrophobic association in dilute aqueous solutions at low solute concentrations (Frank and Evans, 1975) and it may also be that in methanol–water mobile phases the alkyl bonded chains will associate and then lay flat on the surface of the silica to form a hydrocarbonaceous sheath which interacts with the mobile phase (Lochmuller and Wilder, 1979). Under these conditions the stationary phase may be regarded as a quasi-liquid with hydrophobic solutes partitioning into the stationary phase. Such a model would explain the observed changes in selectivity of reversed phase systems with alterations in the chain length of bonded alkyl phases but is complicated by several associated phenomena such as non-uniform distribution of the bonded phase on the silica surface, the rate of penetration of different solutes and the temperature dependence of hydrophobic interactions which might therefore alter the proposed association of the alkyl chains. Consequently, it must be recognised that at present no single, simple model can describe the nature of the surface coverage of bonded phases in reversed phase chromatography. However, a theory has been developed by Horvath and colleagues which allows some prediction of selectivity and retention in reversed phase liquid chromatography and is based upon the characteristics of the mobile phase, giving rise to the term solvophobic theory. The principles underlying this theory will now be described and will be followed by a brief description of its practical application to HPLC separations.

7.3.2. Solvophobic theory

Reversed phase chromatography differs from most other forms of chromatography, which rely upon an attractive force between the solute and the stationary phase. In reversed phase chromatography the stationary phase is inert and only non-polar interactions are possible with the solutes; thus, the essential chromatographic characteristics are governed by solvent effects or, as they are more commonly known, solvophobic interactions. The solvophobic theory was adapted by Horvath and co-workers (Horvath et al., 1976) from the original work on solvophobic theory by Sinanoglu and co-workers (Sinanoglu and Abdulnur, 1965; Halicioglu and Sinanoglu, 1969) and suggests that the energetics of the behaviour of a solute consists of an interaction in the gas phase and a solvent effect. The solvent effect is made up of the positive free energy required to form a cavity for the solute within the solvent and is calculated from the surface tension of the solvent and the surface area of the solute molecules. A positive free energy also arises as a result of a large negative entropy due to bringing a solute molecule from the gas phase into solution and the free energy associated with the reduction in free volume, which is dependent on the density of the solvent.

When the solute molecule dissolves in the solvent two interactions are possible which produce a negative free energy change, these being Van der Waals and electrostatic interactions. The energy associated with the Van der Waals interactions is approximately proportional to the molecular surface area of the solute, while the electrostatic forces are dependent upon the dielectric constant of the solvent and the dipole moment of the solute molecules.

The association process of the solute binding to a ligand of the hydrocarbonaceous bonded phase may be evaluated in a similar fashion; that is, the individual energies associated with each of the processes described above are summed to yield the free energy change associated with the solvent effect. If the interaction of each of the species in the gas phase is known, which in reversed phase chromatography will be primarily Van der Waals forces, the total free energy change of the association process may be evaluated. Thus the binding of a solute to the stationary phase is essentially due to solvent effects with association being facilitated by the decrease in the molecular surface area exposed to the solvent upon complex formation, while attractive interactions with the solvent oppose these interactions. The non-polar interaction between the solute and the hydrocarbonaceous ligand which determines retention results from a difference between these two effects.

Several other models have been proposed to account for retention in reversed phase chromatography; of these two have found some degree of popularity and include the concepts of 'molecular connectivity' and 'interaction indexes'. The value of the molecular connectivity index has been shown to be proportional to the capacity ratio and the solubility of the solute in water (Karger et al., 1976).

The interaction index is somewhat similar to Snyder's polarity index, P, although significant differences are observable with polar compounds (Jandera et al., 1982). The interaction index defines the interactions between the solute and the mobile phase and the model demonstrates that there is a quadratic relationship between the log of the capacity ratio and the volume fraction of organic solvent in the eluent. A consequence of this model is that there is a linear relationship between the corrected log of the capacity ratio; $\log k^* = (\log k' - \log \phi)/V_x$ and the interaction index, where ϕ is the phase ratio and V_x is the molar volume of the solvent. The retention of a specific solute may therefore be predicted from the interaction index of the solute and specific physical parameters of the solvent. This model has been used to accurately determine the retention behaviour of solutes in both binary and ternary solvent systems (Jandera et al., 1982; Colin et al., 1983a).

7.3.3. Stationary phase effects

The application of solvophobic theory to the retention behaviour of solutes has allowed a much clearer understanding of the physicochemical parameters which determine chromatographic behaviour. However, it should be emphasised that solvophobic theory is based on an idealised stationary phase although the nature of this structure remains undefined. The assumptions made in solvophobic theory are reflected in irregular chromatographic behaviour in a number of chromatographic systems for which a number of different explanations have been proposed. Most of the anomalous behaviour can be attributed to solute interactions with free silanol groups at the surface of the stationary phase and has resulted in a modification of the basic solvophobic theory to provide a dual retention model which corrects for silanophilic interactions (Nahum and Horvath, 1981). Such interactions may be masked by either an increase in the water concentration of the mobile phase or incorporation of a suitable amine which could be a buffer component. Conversely, an increase in the methanol concentration of the mobile phase when separating purines and pyrimidines has been shown to enhance a mixed retention mechanism, probably by increasing the surface area of the alkyl chains of the bonded phase and thereby promoting hydrophilic interactions with the stationary phase. Alternatively, the methanol may compete with the solutes for silanophilic interactions, especially in columns with a low surface coverage. The increase in the methanol content may also cause a change in the solvated state of the solutes, thereby altering retention behaviour (Zakaria and Brown, 1983). The strategy of decreasing silanol–solute interactions may improve both chromatographic behaviour and peak shape by preventing tailing of peaks. Alternatively, silanophilic interactions may enhance chromatographic selectivity as can be seen in Fig. 7.1, where masking of the free silanol groups by dodecyltrimethylammonium chloride decreases the resolution of some synthetic peptides.

In addition to solvophobic and silanophilic interactions other stationary phase effects, including steric recognition and pi–pi interactions between solutes and stationary phase, have been demonstrated. Thus, the specific nature of the stationary phase and solutes in-

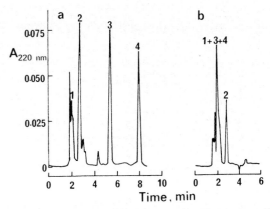

Fig. 7.1. Effect of silanol masking on peptide separation. Chromatographic conditions: column, Supelcosil LC-8 (250 mm×4.6 mm I.D.); mobile phase, acetonitrile–50 mM sodium phosphate in water, pH 2.25 (90:10) without (a) or with (b) 5 mM dodecyltrimethylammonium chloride; flow rate, 1.5 ml/min; temperature, 25°C; detection, UV at 220 nm. Peptide peaks: 1, Boc-Val-Met-Ala-Gly-Val-Ile-Gly-OEt; 2, Boc-Leu-Leu-Ile-Ser(Bzl)-Gly-Oet; 3, NH$_2$-Val-Met-Ala-Gly-Val-Ile-Gly-OEt; 4, NH$_2$-Leu-Ile-Ser(Bzl)Tyr(Bzl)-Gly-OEt. Reproduced from Bij et al. (1981), with permission.

fluences the degree of interaction; for example, planar solutes were shown to be preferentially retained on stationary phases which possessed a planar structure, extended octadecyl groups or large aromatic rings (Tanaka et al., 1982).

7.3.4. Parameters affecting chromatographic retention

7.3.4.1. Organic modifiers. The significance of solvophobic theory is that it is able to predict the retention characteristics of solutes in reversed phase chromatography. This may be clearly demonstrated by the effects of organic modifiers in the mobile phase on the retention behaviour of solutes. In general when pure water is replaced by either pure acetonitrile or pure methanol a linear decrease in the capacity factors of solutes is observed. The explanation for this behaviour is illustrated in Fig. 7.2 where the individual free energy terms constituting the solvent effect are separately shown together with the observed log k'. The Van der Waals term $\Delta G_{\text{VdW}}^{\text{net}}/R \cdot T$ represents the greatest

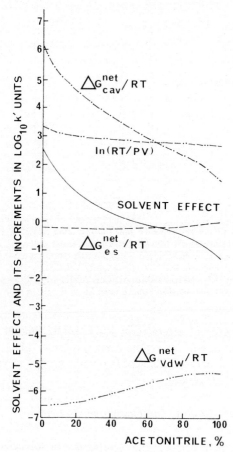

Fig. 7.2. Solvent effects in reversed phase chromatography. The individual terms of the solvophobic equation are plotted as the function of the composition of the water–acetonitrile eluent. The solute is undissociated toluic acid. The data was gathered using an ODS column at ambient temperature. The logarithm of the capacity factor on the ordinate can be considered as a dimensionless energy appropriate at the temperature of the experiment. Reproduced from Horvath and Melander (1978), with permission.

single term, but this, together with the free volume term of the solvent $\ln (R \cdot T / P \cdot V)$ and the electrostatic term $\Delta G_{es}^{net} / R \cdot T$, only cause small changes in $\log k'$. It is thus clear that the major term influencing the solvent effect is $\Delta G_{cav}^{net}/R \cdot T$, representing the change in energy due to cavity formation. Thus, as the acetonitrile concentration is increased the energy required to form a cavity for a solvent molecule is decreased and the tendency for a solute to move into solution is reduced, thereby diminishing the capacity factor. The change in free energy of cavity formation with increasing organic modifier concentration is largely dominated by the surface tension and therefore, as the surface tension of all the commonly used organic modifiers is far less than the extremely high surface tension of water, an increase in the organic modifier concentration will necessarily cause a decrease in the capacity factor.

7.3.4.2. Temperature. The adjustment of column temperature to enhance the speed of chromatographic separations in reversed phase liquid chromatography has become more common in recent years and several specifically designed column heaters are now commercially available. Although separation speeds are usually modified by manipulation of mobile phase composition, an increase in column temperature above ambient levels can provide several additional advantages.

A Van't Hoff plot of ln k' vs. $1/T$ for a series of catecholamines derivatives is shown in Fig 7.3 and demonstrates a straight-line relationship. A general rule of thumb is that a $10°C$ increase in temperature will reduce the capacity factor of the column approximately two-fold (Schmitt et al., 1971).

Deviations from classical Van't Hoff behaviour in reversed phase chromatography have been described (Snyder, 1979; Melander and Horvath, 1979) in which an increased retention of specific chemical types occurs at higher temperatures; this phenomenon may be observed when molecules adopt a compact, near spherical configuration compared with their molecular conformers and are retained for longer times. Also, where the extent of ionisation of either the buffer or the solute molecules is affected by changes in column temperature an altered retention time may be obtained and may result in improved peak symmetry (Knox and Vasvari, 1973; Melander et al., 1979).

Another advantage of chromatography at elevated temperatures is that the rate of interconversion of the alpha and beta anomers of some saccharides may be increased, resulting in the elimination of

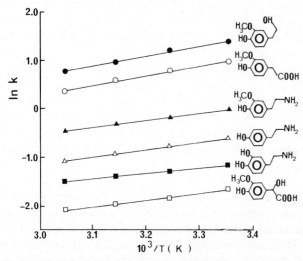

Fig. 7.3. Van't Hoff plots of capacity factors. Chromatographic conditions: column, Partisil 1025 ODS (250 mm × 4.6 mm I.D.); mobile phase, 0.05 M KH_2PO_4; flow rate, 1 ml/min; temperature, 25°C; detection, UV at 254 nm. Reproduced from Horvath et al. (1976), with permission.

unwanted double peaks and a consequent increase in the sensitivity of detection (Vratny et al., 1983).

In addition to increasing the speed of separation by reducing capacity factors, an increase in column temperature will also decrease the viscosity of organic solvents in the mobile phase. This factor may be of considerable importance when the mobile phase has a relatively high viscosity (e.g. methanol–water mixtures) by allowing an increased mobile phase flow rate and a decreased retention time.

7.3.4.3. Added salts. The addition of salts to the mobile phase is a useful strategy to modify the capacity factor of a solute. The way in which salts alter retention behaviour may be conveniently explained by the solvophobic theory, which predicts that for a neutral solute the logarithm of the capacity factor and hence the retention will show a linear increase as the salt concentration increases. This linear increase in retention is a direct consequence of the linear increase in the surface tension.

In the case of ionised solutes, an increase in the salt concentration results in a reduction in the electrostatic repulsion between solute molecules as may be predicted by the modified Debye-Huckel theory of dilute electrolytes (Lietzke et al., 1968). Therefore, for an ionised solute, as the salt concentration is increased, the electrostatic component of the solvophobic equation is increased, resulting in an initial decrease of the ln k' term at low salt concentrations. As the salt concentration is further increased the surface tension effects increase and ln k' increases.

The change in ln k' with salt concentration is shown in Fig. 7.4 for a variety of solutes.

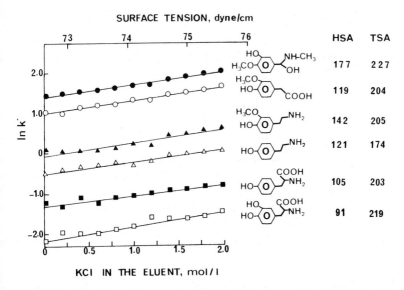

Fig. 7.4. The effect of salt concentration on capacity factor, k'. Chromatographic conditions: column, Partisil 1025 ODS (250 mm×4.6 mm I.D.); mobile phase, the concentration of KCl in a 0.05 M KH_2PO_4 solution was varied; flow rate, 1 ml/min; temperature, 25°C; detection, UV at 254 nm. The upper scale shows the surface tension of the eluent. The hydrocarbonaceous surface area (HSA) and the total surface area (TSA) of the solute molecules are shown. Reproduced from Horvath et al. (1976), with permission.

7.3.4.4. pH. For compounds which ionise when the pH of the mobile phase approaches the pK_a of the solute one can observe major changes in the capacity factor and selectivity. An explanation for this behaviour is provided by the solvophobic theory: ionisation of a solute results in vastly increased electrostatic interactions with the aqueous solvent, resulting in a reduction in retention. The technique of manipulating pH to generate a neutral species whose retention time is increased is termed ion suppression.

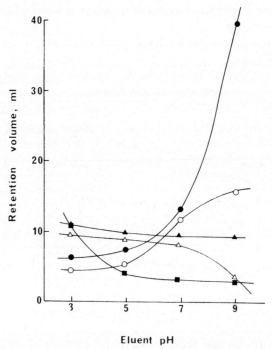

Fig. 7.5. Effect of variation of the eluent pH upon elution volume. Chromatographic conditions: column, μBondapak C_{18} (10 μM) (300 mm$\times$4 mm I.D.); mobile phase, eluent solutions (pH 3.0, 5.0, 7.0 and 9.0) were made up from 0.025 M NaH_2PO_4 and/or 0.025 M Na_2HPO_4 solutions plus 40% methanol (solutions were adjusted to final pH by the addition of 5% sodium hydroxide or phosphoric acid solution); detection, UV at 220 nm. ■, salicylic acid; △, phenobarbitone; ▲, phenacetin; ○, nicotine; ●, methylamphetamine. Reproduced from Twitchett and Moffat (1975), with permission.

The solvophobic theory predicts that, for a solute showing one ionisation for which the equilibrium between the charged and neutral species of the solute is rapid compared with the equilibration with the mobile phase, providing that the binding of each form is determined by intrinsic constants, the capacity factor will vary in a sigmoidal manner with the hydrogen ion concentration; the midpoint of this sigmoid curve will occur where the pH is equal to the pK_a. This model has been extended to solutes where more than one ionisation occurs and the reader is referred to more detailed texts (Horvath et al., 1977a). An example of the chromatographic behaviour of acidic, neutral and basic compounds with varying pH is shown in Fig. 7.5. It may be seen that the weakly basic compounds nicotine and methyl amphetamine show increased retention with increasing pH as they become progressively deprotonated. Phenacetin, a neutral compound, shows no change in retention with changing pH. The weak acids salicylic acid and phenobarbitone show decreased retention with increasing pH. It is clear from this example that, when separating a mixture of solutes whose pK_a's are known, the selection of an appropriate mobile phase pH is essential for the achievement of optimal resolution.

7.3.4.5. Buffers.   In the previous subsection the influence of pH on retention was discussed and it should be emphasised that the pH and buffering capacity of the mobile phase can dramatically influence the peak shape. As we have discussed, ion suppression is a useful technique for increasing the capacity factor of solutes; however, since the degree of solute ionisation can vary with solute concentration in the absence of buffering, as a solute elutes from a column its concentration within the peak varies and so also does the extent of ionisation. Such changes in ionisation can result in tailing or fronting of eluting peaks; where such problems arise, correct buffering of the mobile phase is an easily achievable solution.

In addition to maintaining the pH of the mobile phase, a buffer must also fulfill certain other criteria to be chromatographically useful, including the following.

(*a*) Optical transparency where optical detection systems are being used. It should be noted that acetate, in common with other carboxylic acids, may not be used below 235 nm where it starts to become optically opaque.

(*b*) The buffers must be soluble in the aqueous mobile phase in the presence of the organic modifier.

(*c*) The buffer should be able to mask the free silanol groups in the stationary phase to reduce band spreading.

(*d*) The buffer should play a dynamic role in accelerating the rate of proton equilibration.

7.3.4.6. Chain length. While a number of different stationary phases are available which contain alkyl chains of varying lengths bonded to silica particles, it should be emphasised that the mobile phase com-

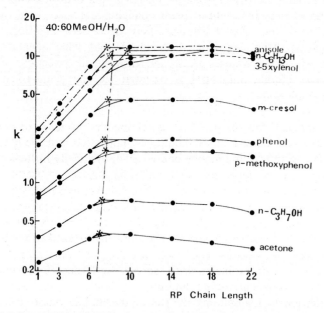

Fig. 7.6. Capacity factor as a function of reversed phase chain length in methanol–water (60:40). Chromatographic conditions: columns, all columns (300 mm × 4.6 mm I.D.) were slurry packed with bonded phases attached to 10 μm SI 100 silica particles; mobile phase, methanol–water (60:40); temperature, 27.5°C; detection, refractive index. Reproduced from Berendsen and De Galan (1980), with permission

position is singly the most dominant factor in determining resolution. However, it would be inappropriate not to consider the effect of carbon chain length on retention. In general, as the length of the bonded alkyl chain is increased, the retention of compounds is increased. Similarly, as the carbon content of the bonded stationary phase is increased, the retention is generally increased. However, there is a limit to the chain length beyond which no improvement occurs; for example, in Fig. 7.6 an increase in the capacity factor is observed up to C_6–C_{10}, beyond which no increase is seen. This has been termed the critical chain length and it is found to vary with different solutes, suggesting that only a portion of the alkyl chain takes part in the separation (Berendsen and De Galan, 1980; Berendsen et al., 1980).

The effect of carbon chain length on selectivity remains a controversial subject and a number of conflicting reports are present in the scientific literature. Thus, examples may be quoted where either an increase or a decrease in the selectivity occurs with increasing chain length depending on the specific solute or mobile phase. However, a general recommendation for reversed phase separations is to use a C_{18} stationary phase, since the retention is greater and the selectivity may be better. Once a separation is defined the length of the alkyl bonded chain may be reduced to provide a reduction in the separation time. However, it must also be recognised that there are exceptions to this generalisation (e.g. protein separations, Section 11.2).

CHAPTER 8

High performance reverse phase ion-pair chromatography

The most serious limitation of conventional reversed phase chromatography is its inability to separate highly polar or ionic compounds from the column void volume whilst retaining good peak symmetry. Until recent years, ion-exchange chromatography was the preferred method for separations involving these types of compounds.

Ion-pair formation by the addition of a counter-ion of opposite charge to the particular solute ion provides a means of charge to the particular solute ion provides a means of charge neutralisation and has been used for several decades in organic chemistry for the extraction of water-soluble ionic compounds into non-aqueous media. The first application of ion-pairing in a chromatographic separation was comparatively recent and has been attributed to Horvath and Lipsky (1966), who used to technique for the separation of charged organic molecules. The emergence of ion-pair chromatography as a popular alternative to ion-exchange chromatography may be attributed to the variety of advantages which the former technique confers, including the following:

(1) Reversed phase ion-pair chromatography may be used to separate both ionic and non-ionised solutes.

(2) The technique is similar to reversed phase chromatography with similar factors affecting the separation; it is therefore easy for the chromatographer who is accustomed to reversed phase techniques to use reversed phase ion-pair chromatography.

(3) The columns used in ion-pair chromatography possess high chromatographic efficiency.

(4) After gradient elutions the re-establishment of original conditions is generally rapid in ion-pair chromatography compared with the rather extended times which may be required with ion-exchange chromatography.

8.1. Modes of ion-pair chromatography

At present, ion-pair chromatography may be divided into three major classes: Normal phase liquid–liquid ion-pair (NPLLIP) chromatography, reversed phase liquid–liquid ion-pair (RPLLIP) chromatography and reversed phase ion-pair (RPIP) chromatography. Each of these chromatographic modes will now be considered, together with their specific advantages and disadvantages.

8.1.1. Normal phase liquid–liquid ion-pair chromatography

In this technique the stationary phase contains the counter-ion and is present in an aqueous coating on the solid support (usually silica particles). The mobile phase is an immiscible non-polar organic solvent often with small amounts of an additional organic modifier. Two major advantages conferred by this mode of chromatography are the ease with which the mobile phase is changed and the fact that specific counter-ions (such as picrate and naphthalene-2-sulphonate) may be added which possess either a strong fluorescent emission or a high extinction coefficient in the UV; the resultant formation of ion-pairs facilitates the measurement of certain otherwise non-detectable solutes. The major disadvantage of this form of ion-pair chromatography is that the preparation of the column is not easily accomplished and the paired ion needs to be repeatedly regenerated on the stationary phase. The technique is not used widely for the simple reason that reversed phase ion-pair chromatography has the convenience of the counter-ion being added to the mobile phase.

8.1.2. Reversed phase liquid–liquid ion-pair chromatography

This technique was introduced in the early 1970s and uses an organic coating around a chemically bonded hydrophobic silica particle support as the stationary phase and an aqueous mobile phase containing the counter-ion. The organic coating utilised is generally pentan-1-ol and the hydrophobic support either octyl or octadecyl bonded phase silica packings. In this mode a chromatographic separation can be manipulated for both selectivity and retention time by adjustment of the concentration of the counter-ion or by alteration of the organic coating comprising the stationary phase.

The advantages of this technique over the normal phase technique are that the columns are relatively stable and the chromatographic conditions can be easily adjusted to improve separations. The disadvantages are that some hydrophobic solute ions have a tendency to partition into the stationary phase, thereby causing a mixed retention mechanism including adsorption on the support in addition to ion-pair formation. Another drawback of this technique is that the organic coating around the hydrophobic support has to be maintained. The obvious solution to overcoming the problems associated with the liquid–liquid chromatographic mode is to use a simple chemically bonded hydrophobic stationary phase and for this reason most efforts today are concentrated on reversed phase ion-pair chromatography.

8.1.3. Reversed phase ion-pair (RPIP) chromatography

RPIP chromatography uses a hydrocarbonaceous stationary phase and either an aqueous or aqueous–organic mobile phase which also contains the counter-ion. The stationary phase is usually an octadecyl bonded phase and the mobile phase is usually an aqueous buffer with either methanol or acetonitrile as an organic modifier. The choice of counter-ions depends on the solutes to be separated, but generally for the separation of acids a hydrophobic organic base is added to the mobile phase, while for the separation of bases a hydrophobic organic acid is added. Separations of other compounds are similarly obtained by the addition of an appropriate counter-ion.

The advantages of this technique over other ion-pair techniques include the high stability and efficiency of the columns and the lack of an organic stationary phase to be regenerated. Moreover, there is a broad range of parameters which may be conveniently adjusted to optimize RPIP chromatography, including the concentration of organic modifier in the mobile phase, the type and concentration of buffer in the mobile phase and the type and concentration of the counter-ion. One final and important advantage is that technically RPIP chromatography is simple and is similar to reversed phase chromatography. The numerous advantages conferred by the RPIP mode of ion-pair chromatography make it the method of choice for many of the applications required by the chromatographer.

8.2. Mechanism of reversed phase ion-pair chromatography

In this section the proposed mechanisms of the chromatographic process will be discussed together with an outline of the influence of the parameters which may be varied on a chromatographic separation.

Throughout its brief history ion-pair chromatography has been known by a variety of names including soap chromatography, solvophobic-ion chromatography, counter-ion chromatography, surfactant chromatography and, more recently, ion-association chromatography. This plethora of nomenclature owes much to the uncertainty which exists concerning the retention mechanism in RPIP chromatography. Five basic models have been proposed in attempts to explain the way in which ion-pairing agents influence chromatographic separations. As the general trend in modern chromatography is to derive and optimise separations on the basis of a known mechanism in a rational manner it is important that each of these models should be considered. However, as all of these models are in some way unsatisfactory in their explanation of the observed chromatographic behaviour and are therefore of rather limited usefulness in optimising separations, only a brief outline of each mechanism will be discussed. For a more detailed discussion of retention mechanisms in ion-pair chromatography the interested reader is referred to excellent reviews by Melander and Horvath (1980a) and Karger et al. (1980).

The first model proposed an ion-exchange mechanism where the free lipophilic charged counter-ion is adsorbed onto the non-polar surface of the stationary phase to constitute an ion-exchange surface. The sample ions of opposite charge to the counter-ions are then partitioned between the stationary phase and the mobile phase by an ion-exchange process. If the concentration or lipophilicity of the counter-ion is increased, the surface coverage of the stationary phase by counter-ion will be increased.

A second mechanism, proposed by Bidlingmeyer et al. (1979) is the so-called ion-interaction mechanism. In this model, the counter-ions are adsorbed onto the surface of the stationary phase to form a primary charged layer, the charge of which is compensated for by the formation of an oppositely charged secondary layer comprised of eluent ions. The retention of sample ions is a function of electrostatic interactions with the charges in the primary and secondary layers and Van der Waals forces associated with adsorption to the surface of the stationary phase. The model allows for sample ions of opposite charge to the counter-ion of the primary layer to penetrate the secondary electrostatic layer in a dynamic manner and then to be adsorbed onto the stationary phase surface. The addition of an ion of opposite charge to the primary layer negates some of the charge and thus allows another counter-ion to be adsorbed onto the stationary phase surface. In this sense an ion-pair is formed at the surface of the stationary phase.

The third mechanism, proposed by Melander and Horvath (1980b), is that of 'dynamic complex exchange' where counter-ions are present both in the mobile phase and at the surface of the stationary phase. Sample molecules which have formed an ion-pair with the counter-ion in the mobile phase can then transfer to the bound counter-ions to form a complex at the surface of the stationary phase. A modification of this process envisages that the complex is initially formed at the stationary phase surface.

The fourth mechanism is that of simple ion-pair formation between the counter-ion and the sample in the mobile phase with subsequent adsorption onto the hydrophobic stationary phase. In this model, in common with those described above, an increase in either the concentration or lipophilicity of the counter-ion would result in an increased retention of solute molecules.

The fifth mechanism, proposed by Stranahan and Deming (1982), utilises a four-parameter thermodynamic model which assumes Langmuir adsorption of the charged counter-ion at the stationary phase–mobile phase interface. The model does not require the formation of ion-pairs between counter-ions and sample ions and assumes that the primary effects of the counter-ion are on the interfacial tension between the adsorbed phase and the bulk liquid phase, together with direct ionic interactions with the charged sample.

The debate concerning the exact mechanism of ion-pair chromatography will no doubt continue. Each of the above models can define the retention behaviour of samples under specific conditions and it may be that the exact mechanism is a combination of several of the abovementioned proposals; indeed, considerable evidence has been accumulated to suggest that the retention mechanism may change from one of ion-pair formation in the mobile phase to one of dynamic ion-exchange as the length of the counter-ion is increased. Thus, in general the ion-pair and ion-exchange mechanisms may be regarded as limiting cases with the true mechanism perhaps being somewhere between the two.

8.3. Choice of experimental parameters

For sample mixtures requiring the use of ion-pair chromatography to establish a good separation, certain choices need to be made to allow the design and subsequent optimisation of the chromatography. Initially, a decision should be made whether reversed phase (RPIP) or normal phase (NPIP) ion-pair chromatography is to be used. As we have previously discussed (Chapter 7), the reversed phase mode provides numerous advantages and with the vast majority of ion-pair separations currently described in the literature using the reversed phase mode, it would appear to be the preferred technique. In subsequent discussions, unless specifically stated, it may therefore be assumed that it is the reversed phase mode which is being described.

After deciding on the mode of separation, the next choice is that of the counter-ion or, as it may sometimes be described, ion-pair reagent. This selection will obviously depend on the nature of the sample, but generally a tertiary amine such as trioctylamine or a quaternary amine such as tetrabutylammonium hydroxide should initially be tried for acids. For basic compounds the usual choice of counter-ion is an alkyl sulphonate (e.g. heptane sulphonate). Variations around the choice and concentration of counter-ion will be described in more detail shortly. The choice of counter-ion, together with the nature of the sample to be separated, requires that a number of other factors be considered in the design of a separation, including mobile phase composition, pH, buffer, stationary phase and operating temperature. Each of the these parameters will now be examined in more detail.

8.3.1. Mobile phase composition

In RPIP chromatography, as in the ordinary reversed phase mode, the mobile phase is usually aqueous with an added organic modifier, which is often methanol or acetonitrile. In NPIP chromatography the mobile phase is usually pentan-1-ol to which hexane or chloroform is added as the modifier. In RPIP chromatography the effect of increasing the methanol or acetonitrile concentration is to reduce retention time (see Fig. 8.1). In this chromatogram, although the elution order of the compounds remains essentially unchanged over the range shown, at higher acetonitrile concentrations an effect on selectivity can occur with the elution of L-DOPA and adrenaline being reversed. Acetonitrile is frequently the preferred organic modifier as its lower viscosity reduces operational pressure; this may be especially important in ion-pair chromatography where the counter-ion reagents are frequently quite viscous. Acetonitrile also has a greater solvation power for quaternary ammonium ions, facilitating the use of higher counter-ion concentrations in the mobile phase.

8.3.2. pH

In each of the specific mechanisms postulated to account for ion-pair chromatography the influence of the pH at which the separation is carried out is absolutely critical. For each separation the pH of the mobile phase should be adjusted to facilitate maximal ionisation of both the sample and the counter-ion; in separations where partial

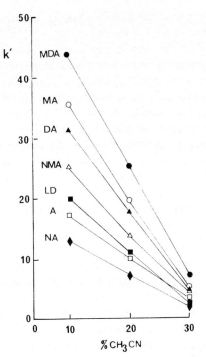

Fig. 8.1. Dependence of k' upon volume percentage of acetonitrile in reversed phase ion-pair chromatography. Chromatographic conditions: column, a polished stainless steel column (125 mm × 5 mm I.D.) packed with ODS/TMS silica (5 μm); mobile phase, water–acetonitrile containing 0.2% (v/v) sulphuric acid and 0.1% (w/v) sodium lauryl sulphate; detection, UV at 280 nm. MDA, 3-O-methyldopamine; MA, metadrenaline; DA, dopamine; NMA, normetadrenaline; LD, L-DOPA; A, adrenaline; NA, noradrenaline. Reproduced from Knox and Jurand (1976), with permission.

ionisation of either sample or counter-ion occurs it generally results in peak broadening. Certain general rules for selection of mobile phase pH may be given:

(a) For samples which are either anions or cations at all pH values, the pH should be selected to allow maximal ionisation of the counter-ion.

(b) For acids the operating pH is usually constrained to the pH range of 7–8.5 and the counter-ion is then selected.

(c) For bases the pH of the mobile phase is usually selected to be in the range of 2–6.

Since the pH of the mobile phase will affect both the capacity factor and the selectivity, it is clear that manipulation of pH may be used to optimise separations of mixtures containing ionogenic compounds. At present nearly all stationary phases employed in reversed phase separations are unstable outside of the pH range of 2–8.5.

A practical limitation on the operation of RPIP chromatography concerns the corrosive activity of many of the counter-ions on the bonded phase of the columns, notably at pH values outside of the range of 2–8.5. It may be anticipated that in the near future the use of non–silica-based stationary phases will result in an increase in the potential applications of ion-pair chromatography.

8.3.3. Effect of added salts

In the previous subsection it was pointed out that the pH of separations in ion-pair chromatography should be carefully controlled; such chromatography therefore requires the addition of buffers which will consequently influence the ionic strength of the mobile phase. In general, in the reversed phase mode an increase in salt concentration causes a decrease in retention, while in normal phase ion-pair chromatography an increase in salt concentration results in an increase in retention. An example of the effect of different salts on the retention of ergotamine in RPIP chromatography is shown in Fig. 8.2. In the chromatogram it is clearly demonstrated that selectivity is influenced by changes in the salt concentration. For certain compounds an increase in the salt concentration may have no effect on retention; this may be due to the opposing effects of: (a) the decreased retention caused by reduced ion-pair formation because of competition by counter-ions and (b) an increase in retention resulting from an increased adsorption of counter-ions on the stationary phase (Knox and Jurand, 1976).

8.3.4. Effect of counter-ion concentration and chain length

In general, in RPIP chromatography an increase in the concentration of the counter-ion results in an increase in sample retention; the

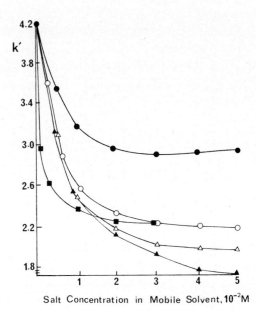

Fig. 8.2. Change of capacity factor (k') of ergotamine with salt concentration in the mobile phase. Chromatographic conditions: column, μBondapak C_{18} (5 μm) (300 mm $\times$ 4.7 mm I.D.); mobile phase, methanol–water (60 : 40) containing 5.0 mM sodium heptanesulphonate and 1.0% acetic acid; flow rate, 2 ml/min; temperature, 20 °C; detection, UV at 254 nm or 245 nm (0.1 aufs). $\bigcirc$, NaCl; $\bullet$, MgSO$_4$; $\triangle$, (NH$_4$)$_2$SO$_4$; $\blacktriangle$, Li$_2$SO$_4$; $\blacksquare$, Na$_2$SO$_4$. Reproduced from Low et al. (1983), with permission.

converse is true for normal phase systems. As shown in Fig. 8.3 the increase in retention with counter-ion concentration is non-linear and a reversal of the trend may occur at higher counter-ion concentrations, an effect which has been proposed to be caused by the counter-ions forming micelles which reduces their effective concentration in the mobile phase. The chromatographer should thus exercise some caution when seeking to enhance sample retention by increasing counter-ion concentration. A change in the selectivity of a chromatographic system may also occur as the counter-ion concentration is increased, although this is dependent upon the specific chromatographic parameters used for each separation. In separations where it is desirable to increase the sample loading it is often possible to achieve

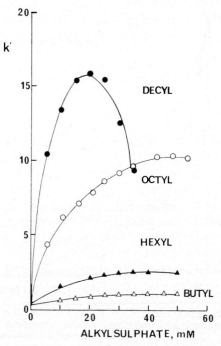

Fig. 8.3. Plots of capacity factor (k') of adrenaline vs. the counter-ion concentration for various *n*-alkylsulphates. Chromatographic conditions: column, Partisil ODS (10 μm); mobile phase, 500 mM potassium phosphate buffer (pH 2.55) containing various concentrations of counter-ion; flow rate, 2 ml/min; temperature, 40 °C; inlet pressure, 400 psi; detection, UV at 254 nm. Reproduced from Horvath et al. (1977b), with permission.

this without adversely affecting the chromatography by increasing the counter-ion concentration.

The effect of an increase in the lipophilicity of the counter-ion used in a RPIP separation is shown in Fig. 8.3. The successive addition of two methylene units in a series of *n*-alkyl sulphates results in large increases in sample retention. The ability to easily manipulate the lipophilicity of the counter-ion in RPIP chromatography provides an extremely convenient method for optimising such separations.

8.3.5. *Effect of temperature*

Retention values in RPIP chromatography are generally slightly more sensitive to changes in column temperature than in the ordinary reversed phase mode but show the same trend of reduced k' values as temperature increases. However, it should be noted that certain separations in ion-pair chromatography are much more sensitive to changes in temperature than others. Thus, it is important that column temperatures are maintained at a constant value and this is easily achieved by the use of a column thermostat. Mobile phases which contain particularly viscous counter-ions may show considerable chromatographic improvements in terms of both speed of separation and peak shape when the temperature of the column is elevated, although under these conditions the life of the column may be slightly reduced depending on the specific conditions employed.

CHAPTER 9

High performance affinity chromatography

9.1. High performance liquid affinity chromatography (HPLAC)

9.1.1. Introduction

The single most important advance in the purification of biomolecules in the last 20 years has been the introduction of affinity chromatography (Cuatrecasas et al., 1968). Affinity chromatography, as discussed

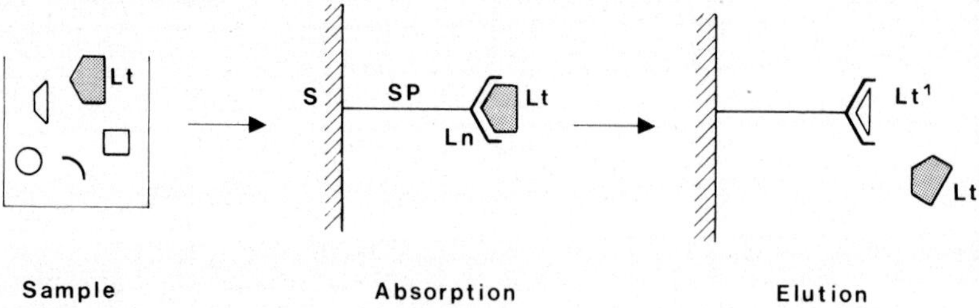

Fig. 9.1. Schematic illustration of the elements involved in bio-affinity chromatography. The solute (Lt) is retained on the stationary phase (S) by specific interaction with the ligand (Ln). The ligand is covalently attached to a spacer arm (SP) which is in turn attached to the stationary phase. Elution as shown here is achieved by specific ligand competition by another solute (Lt^1). Alternatively the ligand–ligate complex can be disrupted by reversible denaturation using low pH solvent or mild chaotropic salts, e.g. sodium thiocyanate.

in this section, refers to biospecific affinity chromatography (Fig. 9.1). An excellent review of affinity chromatography on soft gel matrices can be found in an earlier edition of this series (Lowe, 1979).

9.1.2. Principles

The principles of affinity chromatography have been reviewed in detail recently (Dean et al., 1985) and only an overview will be presented. Affinity chromatography requires that an immobilized ligand (Ln) is covalently attached to the stationary phase and interacts specifically with the solute of interest (ligate, Lt). Elution can be achieved either by specific ligand competition (Fig. 9.1), or by reversible conformational changes of the ligand–ligate complex induced by alteration of the mobile phase parameters. This procedure allows elution of the sample and recovery of activity of the ligand. Binding of the ligate to the ligand is likely to be multivalent for a polymer (i.e. multisite attachment) and may therefore be sterically dependent. The critical features affecting retention are the dissociation constant (K_d) for the ligand–ligate complex and the concentration of the ligand (the number of binding sites) on the stationary phase. The dissociation constant (K_d) is defined by the mass action ratio:

$$[Lt] + [Ln] \rightarrow [LtLn]$$

$$K_d = \frac{[LtLn]}{[Lt] \times [Ln]} \qquad (9.1)$$

The capacity factor (k') for affinity chromatography is defined simply as the ratio of the bound to the free ligate for a given column and can be represented by the equation:

$$k' = \frac{[LtLn]}{[Lt]} \qquad (9.2)$$

A relationship between K_d and k' can be derived when $[Ln] \gg [Lt]$, which is the case in analytical separations. This relationship (Eq. 9.3) has been considered for multivalent attachment (Lowe, 1979; Dean et al., 1985). The dependence of plate height on K_d has been noted using a silica support to which concanavalin A was attached (Muller and Carr, 1984).

$$k' = \frac{[\text{Ln}]}{K_d} \qquad (9.3)$$

9.1.3. Stationary phase effects

9.1.3.1. Chemistry of ligand attachment. A variety of published methods are available for coupling ligands to conventional stationary phases (Fig. 9.2) and most of the stationary phases which are suitable for high performance affinity chromatography have been prepared using some of these methods. In most procedures, silica is coated with a hydrophilic layer (glycophase) to which a reactive amine can then be covalently linked (Fig. 9.3). The most stable supports are generated either by epoxy–silica coupling (Larsson, 1984), or by tresyl-chloride–activated silica (Nilsson and Larsson, 1983).

9.1.3.2. Other characteristics. Developments in particle shape have improved the rate of mass transfer while the increased rigidity of the support has allowed faster flow rates during elution and re-equilibration. It is worth noting that the majority of successful separations have employed wide pore (50–100 μm) silica particles as a base for the stationary phase (Walters, 1982), thus avoiding any steric hindrance. It was found from early work on conventional supports that the presence of a spacer arm was beneficial for binding of the ligate, presumably by allowing freedom of access to the reactive sites of the ligand. However, these arms should not participate in the binding of the ligand and therefore must be both hydrophilic and non-ionic.

It has been estimated that although a high degree of ligand substitution can be achieved on conventional affinity supports, in some cases only 1–2% of the potential sites may be available for binding due to a combination of incorrect orientation and denaturation during coupling (Scopes, 1982). In addition, the molar quantity

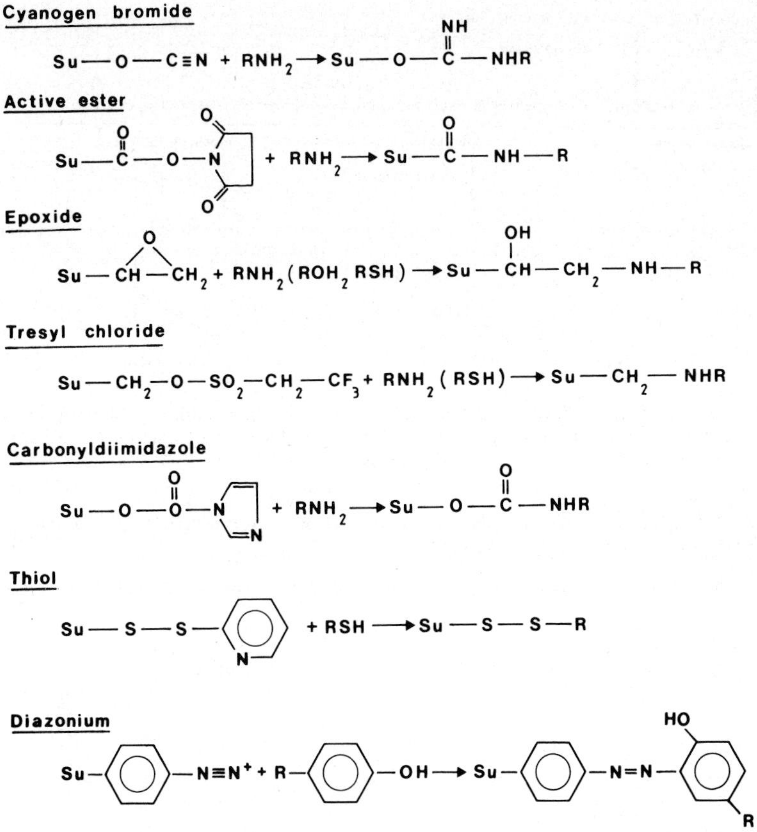

Cyanogen bromide

$$Su-O-C\equiv N + RNH_2 \rightarrow Su-O-\overset{\overset{\displaystyle NH}{\|}}{C}-NHR$$

Active ester

$$Su-\overset{\overset{\displaystyle O}{\|}}{C}-O-N \qquad + RNH_2 \rightarrow Su-\overset{\overset{\displaystyle O}{\|}}{C}-NH-R$$

Epoxide

$$Su-\overset{\overset{\displaystyle O}{\triangle}}{CH}-CH_2 + RNH_2(ROH_2\ RSH) \rightarrow Su-\overset{\overset{\displaystyle OH}{|}}{CH}-CH_2-NH-R$$

Tresyl chloride

$$Su-CH_2-O-SO_2-CH_2-CF_3 + RNH_2\ (RSH) \rightarrow Su-CH_2-NHR$$

Carbonyldiimidazole

$$Su-O-\overset{\overset{\displaystyle O}{\|}}{O}-N \qquad + RNH_2 \rightarrow Su-O-\overset{\overset{\displaystyle O}{\|}}{C}-NHR$$

Thiol

$$Su-S-S- \qquad + RSH \rightarrow Su-S-S-R$$

Diazonium

$$Su- \qquad -N\equiv N^+ + R- \qquad -OH \rightarrow Su- \qquad -N=N- \qquad$$

Fig. 9.2. Some coupling reactions used in the preparation of affinity ligands on conventional low pressure supports (agarose). Most are now available commercially (Pierce Alum, Pharmacia, Miles Yeda).

of ligand that can be coupled to the stationary phase is a function of the molecular size. This means that the capacity of this form of stationary phase can be as much as 100 times less than ion-exchange or reversed phase stationary phases. In cases reported for the preparation of immunoaffinity stationary phases the ligand is usually coupled at a concentration of 5–10 mg/g resin, although in one study alcohol

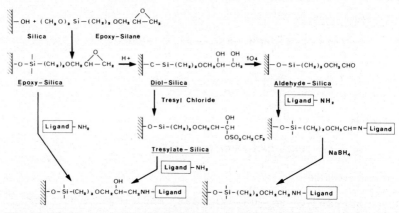

Fig. 9.3. Chemical routes for the preparation of silica-based affinity supports using ligands which contain primary amino groups. Reproduced from Larsson (1984), with permission.

dehydrogenase was coupled to silica at a concentration of 21 mg/g (Nilsson and Larsson, 1983).

A number of studies have used textile dyes as affinity ligands,

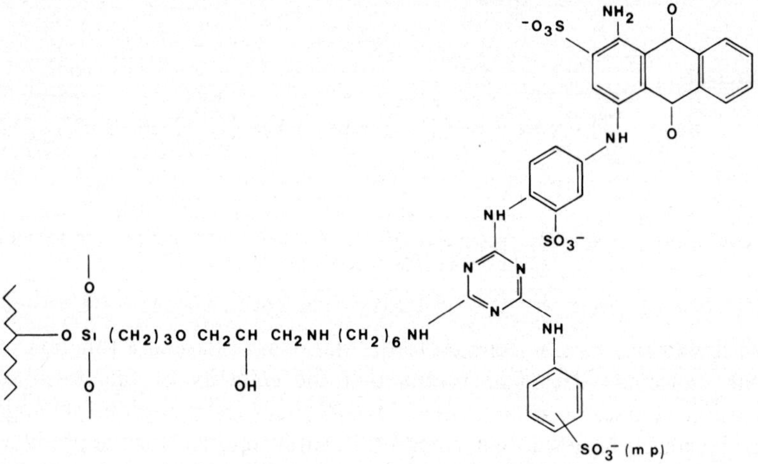

Fig. 9.4. Structure of Cibacron Blue FG3-A. The dye is shown coupled to silica via a spacer arm. Reproduced from Lowe et al. (1981).

notably Cibacron Blue FG3-A (Fig. 9.4). This molecule contains a sulphonated anthraquinone moiety which is thought to mimic many of the nucleotide cofactors required for catalytic activity of enzymes (Lowe et al., 1981). A dramatic demonstration of the power of resolution of this type of support was demonstrated for the two enzymes, hexokinase and 3-phosphoglycerate kinase, which were isolated from crude yeast extracts by selective elution using their respective substrates.

Immunoaffinity. The majority of polyclonal antibody preparations have dissociation constants in the order of $K_d = 10^{-8}$ to 10^{-12} M. Whereas this allows for maximum selectivity, the ligand–ligate dissociation can become rate limiting during chromatography, hence their use in HPLAC, where speed is required, may not be considered advantageous (Horvath and Lin, 1978). Flow rates in the order of 0.05–0.1 ml/min may be required to allow successful chromatography and this may defeat the purpose of coupling antibody ligands to high performance supports. The possibility of selecting monoclonal antibodies with lower binding constants may offer a preferable alternative (Sportsman and Wilson, 1980). Table 9.1 shows the more

TABLE 9.1
Commercially available affinity ligands and supports

Class	Ligand	Ligate
Biospecific	Enzyme	Cofactor
		Substrate
		Inhibitor
	Antibody	Antigen
Group-specific	Protein A	IGG
	Concanavalin A	D-Sugars
	Lectins	Carbohydrates
	Poly U	Nucleic acids
	Poly A	Nucleic acids
	Poly Lys	Nucleic acids
	Triazine dye	Nucleotides
	Nucleoside	Nucleotides
	Boronic acid	1,2-*cis*-Diol sugars
	Thiol	Cysteine

popular types of affinity ligands which are available commercially divided into two classes: those that are recognised as being strictly biospecific and those that separate molecules that share a structural or functional homology. A comprehensive list of affinity ligands has been presented elsewhere (Lowe, 1979) but the list is continually growing.

9.1.4. Mobile phase effects

Selectivity is ultimately determined by the specificity of the ligand–ligate interaction. Although the binding may largely be governed by stereospecific interactions, K_d will be dependent on electrostatic, hydrophobic and other such interactions and therefore will be influenced by mobile phase characteristics such as pH and ionic strength. Kinetic studies of the ligand–ligate interaction have shown conflicting results which may be due to the different methods used to couple the ligand or to leakage of the ligand from the support. (The leakage of the ligand from the support has been a drawback in the use of large scale affinity supports for the production of pharmaceutically active compounds.) Solution studies on K_d indicate that as the temperature is increased there is an increase in K_d, and a corresponding decrease in k'.

9.2. High performance ligand exchange chromatography (HPLEC)

9.2.1. Introduction

The development of high performance ligand exchange chromatography (HPLEC) has been quite distinct from HPLAC. Much of the impetus arose from attempts to separate racemic forms of amino acids (Rogozhin and Davankov, 1971). One approach was to convert the amino acids to diastereomeric mixtures by modification of the amino or carboxyl group, followed by normal phase chromatography. Popular derivatising agents are isothiocyanate and thiourea compounds (Furakawa et al., 1977; Namambara et al., 1974). Although still popular, this approach suffers the disadvantage of the long derivatisation and the possible generation of artifacts during the reaction. Another approach has been to add a chiral agent to the mobile phase during either normal or reversed phase chromatography (Lindner et al., 1980). More recently, increasing use has been made of stationary phases to which a chiral agent has been covalently attached. The following sections (9.2.2, 9.2.3 and 9.2.3) will deal with these.

9.2.2. Principles

The most popular mode of HPLEC is that sometimes referred to as metal chelate chromatography (Davankov and Semechkin, 1977; Hemdan and Porath, 1985a, 1985b, 1985c). In HPLEC the retention of a solute (Lt) is dependent on the formation of a metal ion–ligate complex. The metal ion is itself bound in a coordination complex formed with the stationary phase (SM). The metal ion–ligate complex (SMLt) of the stationary phase is reversible and has a highly ordered structure. This structure imparts the property of stereospecificity, allowing the separation of compounds which possess a very similar structure (e.g. geometric and optical isomers). In this chromatographic mode the factors which influence retention are quite different from those in ion-exchange chromatography and this is reflected in the high salt concentration which is utilised in the equilibration buffer. The chromatographic process can be represented as follows:

Equilibration: $SH^+ + M \rightarrow SM + H^+$
Retention: $SM + Lt \rightarrow SMLt$
Elution: $SMLt + 2H^+ \rightarrow SH + M^+ + LtH^+$

The stationary phase is initially equilibrated as the protonated form (SH^+), the proton is then displaced by the metal ion (M) and subsequently the ligate is coordinated (SMLt). Elution is achieved by acidification or by the addition of a competing ligate (not shown), although the latter does not remove the metal ion.

9.2.3. Stationary phase characteristics

Metal chelate chromatography was originally developed using low pressure stationary phases such as agarose (Hemdan and Porath,

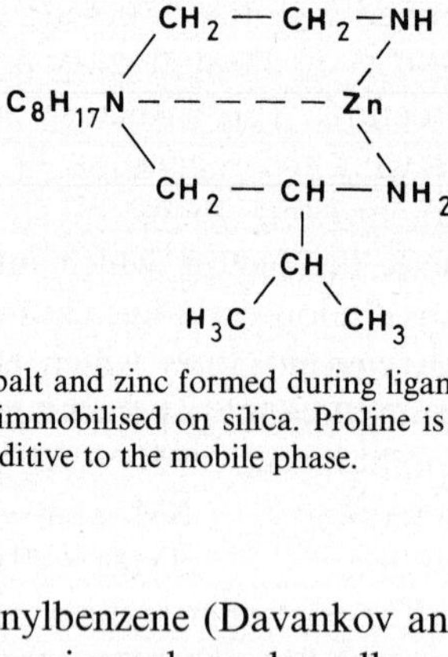

Fig. 9.5. Some coordination complexes of copper, cobalt and zinc formed during ligand exchange chromatography. 1 and 3 represent those immobilised on silica. Proline is a popular ligand and is also used as a chiral additive to the mobile phase.

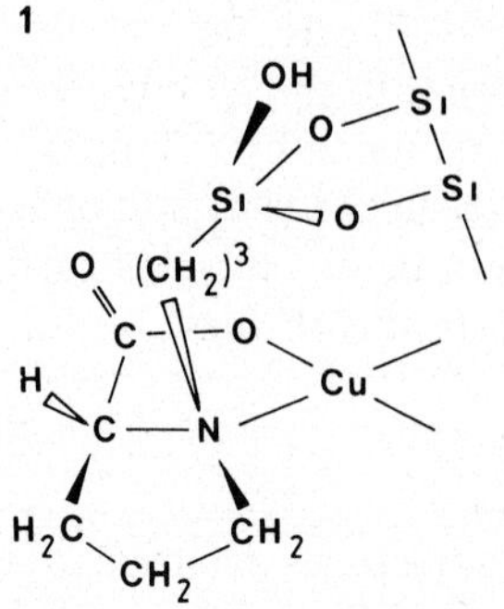

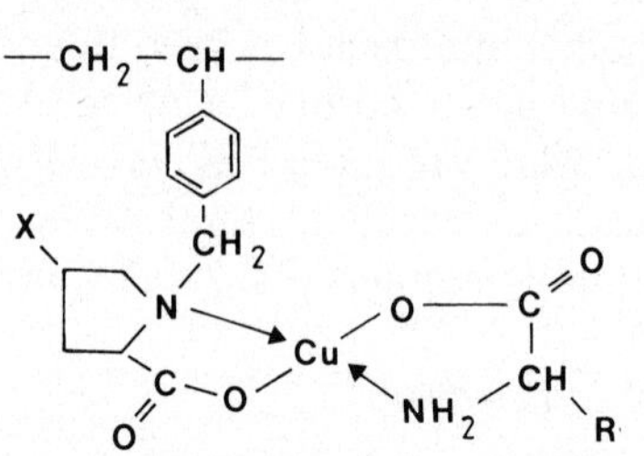

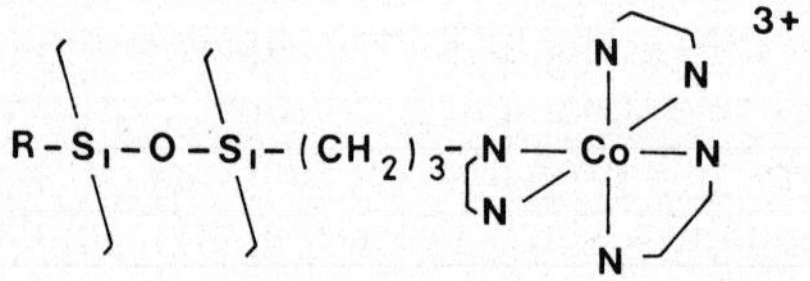

1985a, 1985b, 1985c) and polystyrene divinylbenzene (Davankov and Zolotarev, 1978). The introduction of silica microspheres has allowed the use of faster flow rates and resulted in improved efficiency (Kurganov et al., 1983). Some examples of silica-bound metal ion complexes are shown in Fig. 9.5 (1 and 3). A popular chiral ligand which is used in HPLEC is the amino acid proline. Alternatively, the addition of a proline–copper complex to the mobile phase was used to separate D,L-amino acids on an ion-exchange stationary phase (Hare and Gil-Av, 1979). HPLEC stationary phases are available commercially which contain both small chiral molecules (Diacel Chiralpak W)

TABLE 9.2

A list of compounds separated on Enatiopac (silica–glycoprotein)
(reproduced from LKB technical brochure, with permission)

Substance	α	Substance	α
Atropine	1.64	Mepivicane	1.25
Bromodiphen-		Methadone	1.59
hydramine	1.17	Methdorphan	2.54
Brompheniramine	1.50	Methylatropine	1.27
Bupivacaine	1.41	Methylhomatropine	4.2
Butorphanol	1.99	Methylphenidate	1.70
Carbinoxamine	1.33	Metoprolol	1.64
Chlopheniramine	2.66	Nadolol A	3.98
Clidinium	1.21	Nadolol B	3.03
Cocaine	1.46	Oxyphencyclimine	1.42
Cyclopentolate	3.86	Oxprenolol	1.25
Dimethindene	1.53	Phenmetrazine	1.57
Diperodone	1.47	Phenoxybenzamine	1.37
Disopyramide	2.70	Promethazine	1.25
Doxylamine	1.37	Pronethalol	1.26
Ephedrine	1.83	Propoxyphene	2.3
Ephedrine (pseudo-)	1.34	Propranolol	1.13
Homatropine	1.63	Terbutaline	1.22
Labetalol A	2.10	Tocaine	1.44
Labetalol B	1.36	Tridihexethyl	1.64
Mepensolate	1.32	Trimeprazine	1.11

and larger biopolymers (LKB Enantiopak). A list of compounds separated using the latter stationary phase is shown in Table 9.2.

The methods which are used for the preparation of HPLEC stationary phases are similar to those which have been developed for soft gel matrices: by treating controlled pore glass with 3-(2-aminoethylamino)propyl-trimethoxysilane and subsequently perfusing the column with copper sulphate, a copper-chelate support is prepared (Masters and Leyden, 1978). Direct treatment of silica with copper sulphate can also be used (Caude and Foucault, 1979) (Fig. 9.6). Methods used in the preparation of these stationary phases can be found in the literature (Sugden et al., 1980; Caude et al., 1984).

9.2.4. Mobile phase effects

The use of mobile phase chiral additives for the separation of amino acids and dansyl amino acids has been reviewed previously (Gil-Av

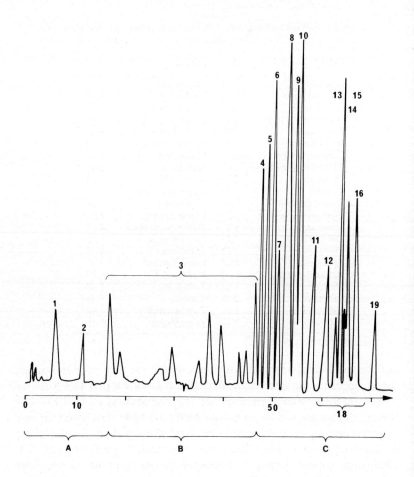

Fig. 9.6. Composite diagram showing the elution position of some peptides and amino acids. Chromatographic conditions: column, Lichrosorb Si 60 7 μm (treated with 0.1 M copper(II) sulphate/1 M ammonia); mobile phase, (A) water/acetonitrile (10:90)–0.1 M ammonia, 1 ppm Cu^{2+}, (B) water–acetonitrile (60:40)–0.95 M ammonia, 1 ppm Cu^{2+}. Elution was achieved with a concave gradient of 0% B to 100% B over 70 min; flow rate, 2 ml/min; detection, UV at 254 nm. 1, Phe-Phe; 2, Ala-Ala-Ala; 3, mixture; 4, Ala-Ser; 5, Pro-Glu; 6, Phe; 7, Gly-Gly-Gly; 8, Lys-Phe; 9, Leu; 10, Leu; 11, Glu; 12, Ala; 13, Ser-Ser-Ser; 14, Gly-His-Gly; 15, Arg-Glu; 16, Lys-Gly; 17, Arg-Tyr; 18, Pro-Gly-Lys-Ala-Arg,Lys-Lys-Gly-Glu; A, hydrophobic, large peptides; B, dipeptides; C, amino acids, hydrophilic peptides, basic peptides.

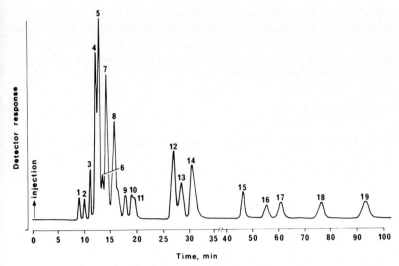

Fig. 9.7. Separation of a mixture of enantiomeric amino acids on a reversed phase column with a chiral agent added to the mobile phase. Chromatographic conditions: column, Spherisorb/LC-18 (5 μm, 150×4.6 mm I.D.); mobile phase, water; chiral additive, Cu(N, N-di-n-propyl-L-alanine); flow rate, 0.2 ml/min; temperature, ambient. 1, D-lysine; 2, L-lysine; 3, D-arginine; 4, glycine; 5, D-alanine, D-serine; 6, D-threonine; 7, D-arginine; 8, L-threonine, L-alanine, D-histidine; 9, D-cysteic acid; 10, L-cysteic acid; 11, L-histidine; 12, D-aspartic acid; 13, D-valine; 14, L-aspartic acid; 15, D-glutamic acid; 16, D-methionine; 17, L-valine; 18, L-glutamic acid; 19, L-methionine.

and Weinstein, 1984). Most separations have been performed in the reversed phase mode using a ligand such as N, N-di-n-propyl-1-alanine with copper(II) as a metal ion (Fig. 9.7). Those species that are not initially resolved can be separated by alteration of the mobile phase conditions. Since the formation of the coordination complex is dependent on dipole interactions, alteration of mobile phase parameters such as pH, ionic strength, temperature, and also the type of metal ion will affect the retention. The most common metal ions used in HPLEC are: Zn(II), Cu(II), Ni(II), Co(II), Hg(II) and Cd(II) (Fig. 9.5).

In a study of the retention of amino acids on immobilised nickel iminodiacetate it was found that variation of the pH has a differential effect on different amino acids due to the ionisation of both the amino acids and the stationary phase to which the metal ion is coordinated. The residues histidine and cysteine will form stable complexes with Zn(II) and Cu(II) at neutral pH and therefore are the critical residues when separating either peptides or proteins (Hemdan and Porath, 1985a, 1985b, 1985c).

9.3. Summary

Both HPLAC and HPLEC are new introductions to the field of chromatography and currently relatively few of the commercially available stationary phases are widely used. The most significant contribution made by HPLEC is in the simplification of the purification of optical isomers and its potential use in the separation of other geometric isomers.

CHAPTER 10

Practical aspects

10.1. Introduction

The purpose of this chapter is to familiarise or remind the chromatographer of the more general considerations which are vital for successful and reproducible separations. In doing so, the chapter contains a

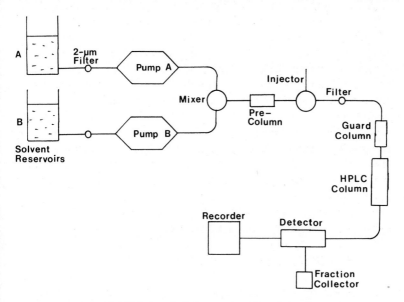

Fig. 10.1. A typical chromatographic system.

number of chromatographic hints and tips which should benefit both the novice and the more experienced chromatographer. The chromatographic system will be discussed as a whole (Fig. 10.1) and later the individual components will be examined.

10.2. General operation

Initially, the solvent reservoirs should be checked to ensure that they contain sufficient mobile phase for the proposed chromatographic separation and so prevent air being drawn into the pumps. The mobile phase can then be pumped through the column until a steady baseline signal is obtained from the detector. This would usually entail a flow rate of between 0.5 and 4 ml/min for a 4.6 mm I.D. column for anything between 5 and 60 min. Care should be taken to ensure that the column back pressure does not exceed the recommended limitations for the column as this could cause irreversible damage.

A test sample should then be injected to ensure that the system is functioning properly and that sufficient selectivity, resolution and retention is obtained for the particular application. It is important to ensure that each of the components which were loaded onto the column are subsequently eluted from the column. This is achieved either by knowledge of the sample composition or by flushing the column with a much stronger mobile phase to remove any strongly bound material. Care should always be taken to ensure that the mobile phase is compatible with the column. In the event that components remain on the column these may subsequently slowly leach off, causing either aberrant peaks or excessive baseline noise in the following chromatographic run. In addition, impurities may accumulate on the column, causing a reduction in column efficiency. The chromatography may be repeated with alternative mobile phases until satisfactory chromatograms are obtained.

10.3. Establishing an HPLC separation

The method by which a successful HPLC analysis is established is largely determined by what is known about the sample composition.

For example, for known compounds of a specific type the reader would be recommended to refer to the applications sections contained within this book to ascertain which systems have been previously described. Alternatively, for an unknown compound or for one not described in this book, the reader will be able to establish a separation based on the first principles outlined in this chapter and Chapter 2.

10.3.1. Mode selection

For a given sample little information is required to allow the selection of the most appropriate stationary phase. Selection may easily be made using the mode selection chart in Table 10.1. It can be seen that the mode appropriate to any particular application depends partly upon the molecular weight of the sample and partly its physicochemical character. Essentially there are four main categories of separation, which are briefly outlined below.

TABLE 10.1
Mode selection

Sample			Mode
M.W. < 2,000	Non-ionic	High polarity	Normal phase
		Low polarity	Reversed phase
	Ionic	Acidic	Anion-exchange
			Reversed phase ion-pair
		Basic	Cation-exchange
			Reversed phase ion-pair
M.W. > 2,000	Water soluble	Ionic	Ion-exchange
		Polar	Normal phase
		Non-ionic	Reversed phase
			Affinity
			Ligand exchange
			Normal phase
			Size exclusion with aqueous mobile phase
	Water insoluble		Size exclusion with organic mobile phase
			Normal phase in organic solvents

10.3.1.1. Size exclusion chromatography. This mode relies on the accessibility of the pores in the stationary phase to molecules of differing molecular sizes and has particular application to high molecular weight biological samples such as proteins, carbohydrates and lipids. Smaller components are able to penetrate further into the pores and are therefore retained longer on the column than larger molecules, which are excluded from the pores.

10.3.1.2. Adsorption chromatography. In this mode, silica (or sometimes alumina) is usually employed as the stationary phase although this method is not commonly used for biological samples and is often replaced by bonded normal phase stationary phases. The method relies on the differing affinities of the sample for the mobile phase and the stationary phase.

10.3.1.3. Partition chromatography. Up to 90% of all applications use this mode of chromatography which relies on the partition of solutes between two immiscible liquid phases; one bound to a solid support and the other mobile. The most popular stationary phase is an octadecylsilane (ODS or C_{18})-coated silica particle, although lower homologues (C_8, C_2) are also employed. The popularity of reversed phase chromatography is increased by the availability of a large number of commercial packings varying in chain length, chain type and carbon loading. Thus, the chromatographer not only has the choice of varying selectivity by changing mobile phase composition but also of selecting a different stationary phase. It is possible to separate both ionised and non-ionised solutes using reversed phase chromatography. Ionic charges can be suppressed by manipulation of pH or by the inclusion of an ion-pair reagent.

For more polar components, the bonded normal phase supports are particularly useful and have benefits over adsorption chromatography on silica by allowing a rapid response to solvent composition changes. Other modes of partition chromatography, such as ligand-exchange and affinity chromatography, have found particular applications for the resolution of water-soluble biomolecules.

10.3.1.4. Ion-exchange chromatography. Retention in ion-exchange chromatography is governed by electrostatic interactions between oppositely charged groups of the solute and the stationary phase. Selectivity can be affected by the nature of the mobile phase or the temperature. Certain applications, particularly for multicomponent samples, require more than one HPLC mode to adequately separate the components and this can be achieved using multidimensional HPLC (Majors, 1980).

10.4. Solvent

10.4.1. Solvent selection

This section is designed to give guidance to mobile phase selection from first principles. The choice of mobile phase composition is critical because of the widely differing selectivities which can be obtained. In certain cases it may be that trivial factors such as availability and expense govern the final selection.

There are essentially three criteria for the selection of a suitable mobile phase. These are based on physicochemical properties of the sample, physicochemical properties of the solvents and mobility of the sample by TLC (where possible).

10.4.1.1. Physicochemical properties of the sample. Unusual chromatographic effects can be observed if the sample is applied to the column in a solvent other than that being employed for the separation. If, for example, the sample is dissolved in a stronger solvent, then a band in which column equilibration is disturbed will progress down the column and will often impair resolution. Therefore, as a general rule it is always best to choose a mobile phase in which the sample is soluble and to use a larger injection volume if solubility is a limiting factor.

10.4.1.2. Physicochemical properties of the solvents. Table 10.2 shows the common physicochemical properties of various solvents. In theory any of these solvents could be used for chromatography, but for

TABLE 10.2

Physicochemical properties of various solvents

Solvent	Density	BP	RI (n_D^{20})	Viscosity (cP, 20 °C)	Polarity ($e°(Al_2O_3)$)	UV cutoff (nm)
Fluoroalkanes					−0.25	
n-Pentane	0.626	36.0	1.358	0.23	0.00	210
2,2,4-Trimethylpentane	0.692	98.5	1.392		0.01	210
Hexane	0.659	86.2	1.375	0.33	0.01	200
Cyclohexane	0.779	81.4	1.427	1.00	0.04	210
Carbon tetrachloride	1.590	76.8	1.466	0.97	0.18	265
Toluene	0.867	110.6	1.497	0.59	0.29	285
Benzene	0.874	80.0	1.501	0.65	0.32	280
Diethyl ether	0.713	34.6	1.353	0.23	0.38	220
Chloroform	1.500	61.2	1.443	0.57	0.40	245
Methylene chloride	1.336	40.1	1.424	0.44	0.42	240
Tetrahydrofuran	0.880	66.0	1.408	0.55	0.45	215
Methylethylketone	0.805	80.0	1.378		0.51	330
Acetone	0.818	56.5	1.359	0.32	0.56	330
Dioxane	1.033	101.3	1.422	1.54	0.56	220
Ethylacetate	0.901	77.2	1.370	0.46	0.58	260
Triethylamine	0.728	89.5	1.401	0.38	0.63	
Acetonitrile	0.782	82.0	1.344	0.37	0.65	200
Pyridine	0.978	115.0	1.510	0.94	0.71	305
n-Propanol	0.804	97.0	1.380	2.30	0.82	210
iso-Propanol	0.785	82.4	1.377	2.30	0.82	210
Ethanol	0.789	78.5	1.361	1.20	0.88	210
Methanol	0.796	64.7	1.329	0.60	0.95	205
Water	1.000	100.0	1.330	1.00	High	
Acetic acid	1.049	117.9	1.372	1.26	High	

practical purposes a number of solvents can usually be eliminated for a variety of reasons:

(a) the solvents with high viscosities cause higher back pressures during flow in the chromatographic column,

(b) the solvents with low boiling points can be difficult to maintain at a fixed concentration in the mobile phase and in addition can present a fire hazard,

(c) solvents with a high UV cut-off wavelength preclude UV detection at commonly used wavelength ranges,

(d) some solvents are toxic (e.g. benzene), and

(e) the purity of some solvents may vary from batch to batch.

Thus, only a limited number of solvents have general utility (Thomas et al., 1979). Acetic acid and triethylamine have been included in Table 10.2 since they are sometimes used in the preparation of aqueous buffers used in ion-exchange chromatography.

Finally, it should be noted that although it is solvent strength which primarily controls retention, hydrogen bonding in protic solvents may also influence the retention.

10.4.1.3. Mobility of the sample by TLC.   Frequently a single solvent will not allow optimal chromatographic resolution and the majority of separations use a mixture of solvents as the mobile phase. The optimum choice of mobile phase can often be determined by trial and error, but some guidance can be obtained using TLC. Theoretically, mobility on a TLC plate is directly related to HPLC retention since $k' = (1 - R_f)/R_f$ (Geiss and Schlitt, 1976); however, in practical terms there are differences since TLC is a non-equilibrium process and generally lower solvent strengths are required for an equivalent HPLC separation. Nevertheless, TLC mobility provides a rapid method for comparing many solvent mixtures.

In general, in the development of a separation of a sample mixture, the sample is initially injected onto the HPLC column in a solvent of low elution strength which is then steadily increased until the compounds of interest elute from the column. The main limitation of this method is that it may take a long time to effect elution and consequently it is often quicker to use this procedure in reverse, with discrete jumps in eluent strength (Parris, 1978). Modern equipment allows gradient development to be carried out unattended.

Many manufacturers advise avoiding the use of halide ions at low pH with stainless steel pumps because of the possibility of corrosion. It is therefore recommended that where halide ions are used they should be flushed from the system with water when it is not in use. When using halide ions it is also recommended that the pump should be passified by pumping a solution of 20% nitric acid through the system (not the column!) for a period of 20 min at least once every week.

10.4.2. Solvent purity

The purity of the mobile phase is a crucial factor in obtaining good HPLC separations, especially where an analysis requires high sensitivities of detection. Impurities in solvents can arise from either dissolved liquids or gases or from suspended particles. These impurities may adsorb to the surface of the column packing and cause a change in its selectivity or, alternatively, they may elute as a single peak with increasing eluent strength. Fortunately, a number of companies market solvents specially distilled or purified for HPLC which are largely free of dissolved impurities. Typical impurities include lower or higher homologues, acids and bases, UV-absorbing compounds and water. For analytical separations laboratory deionised water is generally not sufficiently pure. Methods to more extensively purify water have been published (Sampson, 1977) and HPLC-grade water is now commercially available (Rathburn Chemicals). The reader is directed to other sources for a detailed guide to solvent purification (Bristol, 1980; Rabel, 1980).

Dissolved gases are a particular problem in gradient systems where the solubility of the gas may be less in the mixture than in the single solvent; this may be a significant problem where water and methanol or acetonitrile are mixed because of the exothermic breakage of hydrogen bonds in water which can result in volatilisation of the organic solvent. However, premixing a small percentage of methanol in the aqueous eluent will minimise this problem. In low pressure HPLC systems degassing in the mixing chamber can result in bubbles of gas being pumped onto the chromatographic column with a consequent loss in column performance. However, since the HPLC column is under pressure while in use, degassing usually arises in the tubing on the outlet (low pressure) side of the column, giving rise to gas bubbles in the detector cell which can then create excessive signal noise. In general, degassing procedures rely on; (*a*) boiling, (*b*) vacuum degassing, (*c*) agitation by sonication (often under vacuum), and (*d*) agitation by sparging with nitrogen, argon or helium (these have a low solubility themselves but are expensive). It should be noted that these procedures may change the composition of the liquid and should therefore be carried out prior to mixing. Degassed solvents absorb air very slowly and can be used without special precautions.

Finally, if solid particles in the solvents are drawn into the HPLC system they can impair the sealing of pump valves or clog the top of the chromatographic column. Both pre- and post-pump filtration is therefore strongly recommended. Solvents should be filtered through a 5 μm filter (solvent compatible) into the solvent reservoir. The pre-pump filter should be a 5 μm sintered filter attached to the tube leading to the pump. It is important that this should have a large surface area to facilitate a free flow of solvent, especially with reciprocating pumps. The post-pump filter provides extra protection by preventing the accumulation of fine debris from the pump head onto the column.

10.5. Isocratic vs. gradient elution

Isocratic HPLC systems utilise a single mobile phase and are often preferred on the basis of cost, convenience and improved detector responses. This is especially true for refractive index or electrochemical detectors where the use of gradient systems is usually precluded. Additionally, isocratic elutions do not require column re-equilibration after each run, providing savings both in time and solvent.

The main advantages of gradient over isocratic systems are provided in the analysis of unknown or multicomponent samples where a range of solvent strengths may be required both to ensure that all the components loaded are eluted and also that the retention times of later eluting peaks are minimised. Gradient systems may be subdivided into two classes; those where solvents are mixed prior to the pump (low pressure systems) and those where solvents are mixed after the pump (high pressure systems).

10.5.1. Low pressure systems

In low pressure systems the gradient is generated prior to the pump and is delivered to the column in pulses determined by the dead volume of the pump head. Their main advantage is the relatively low cost. A simple system designed to generate a low pressure gradient has

been described (Perrett, 1976) but does not always give good reproducibility. Commercial systems rely on switching valves but inaccuracies can occur when less than 5% of one of the components is being
delivered.

10.5.2. High pressure systems

High pressure gradient systems employ two pumps whose output is
usually controlled by a microprocessor. These systems usually require
a mixing chamber which allows accurate and reproducible gradients
to be generated. It should be noted that complete mixing of the
mobile phases is important to minimise excessive detector noise.

10.6. Sample preparation

The direct injection of a crude biological sample onto an HPLC
column can cause major problems and it is usually essential that some
form of sample clean-up be undertaken prior to analysis. Failure to
partially purify the sample can allow particulate matter to enter the
HPLC system and also allow soluble components to irreversibly bind
to the stationary phase. An initial clean-up and extraction step often
involves sample homogenisation, although certain compounds which
are bound to proteins or nucleic acids may require either acid or base
hydrolysis to allow their extraction. Enzymatic hydrolysis provides a
milder but slower alternative. Hydrolysis can be combined with
protein denaturation and precipitation using perchloric acid to remove unwanted proteins in a biological sample. Alternatively, protein
can be removed by the addition of other acids, salts (e.g. ammonium
sulphate), solvents or by ultrafiltration through a semi-permeable
membrane. At this stage, providing that the compounds of interest are
present in sufficient concentration, they can be injected onto the
HPLC column. Otherwise some form of concentration by extraction is
necessary and this can be carried out either directly by a reduction of
sample volume through evaporation (e.g. nitrogen or vacuum) or
solvent extraction of the more lipophilic components in the mixture,
or indirectly by first adjusting the pH to suppress ionisation of the

relevant compounds and then extracting them into an organic layer. Neutralisation and back-extraction into an aqueous phase can then be carried out. Alternatively, compounds can be extracted and concentrated by column chromatography for which special columns have now been designed (e.g. Sep Pak; Waters Assoc.). It is important that excess salts used during extraction procedures are removed to maintain column life.

As a final step, the sample should be dissolved in an appropriate solvent (preferably the HPLC mobile phase) and either centrifuged or filtered prior to injection onto the HPLC column. If the concentration of the compounds of interest is insufficient to allow detection the sample may require derivatisation.

10.7. Derivatisation

Derivatisation can be carried out before or after chromatography (Lawrence and Frei, 1976).

10.7.1. Pre-column derivatisation

The advantages of pre-column derivatisation are not only that sample detection can be enhanced but also that extraction, purification and chromatography of the compounds can be modified. In addition, a large choice of reagents are available (Table 10.3). However, the disadvantages of pre-column derivatisation are that samples may degrade during the derivatisation procedure and that the chemical reactions may not go to completion, resulting in spurious peaks in the chromatogram.

10.7.2. Post-column derivatisation

Unlike pre-column derivatisation, the post-column derivatisation reaction need not reach completion, although the extent of reaction must be reproducible (Gfeller et al., 1977). It is this requirement, coupled with the desire to limit band spreading, which has led to reactor design being a crucial factor (Schwedt, 1979). Three reactor

TABLE 10.3
Examples of pre-column derivatisation agents
(reproduced from Bristow (1976),
with permission)

Reagent	Solute type	Detection
Dansyl chloride	Amino acids, phenols amines, peptides	Fluorimetry (ex,340; em,510)
Phenacyl bromide	Fatty acids, phospholipids prostaglandins	UV
Benzoyl chloride	Carbohydrates	UV
3,5-Dinitrobenzoyl chloride	Alcohols, phenols, amines	UV (254 nm)
Fluorescamine	Amino acids, peptides, primary amines	Fluorimetry (ex,395; em,490)
4-Chloro-7-nitrobenzo-2,1,3-oxadiazole	Glycosides, aglycones	UV (260 nm)
Phenylhydrazine	Ketosteroids	UV
Dansylhydrazine	Carbonyl compounds	Fluorimetry
N,N'-Dicyclohexyl-carbodiimide (DCC)	Fatty acids	UV
o-Phthalaldehyde	Amines	Fluorimetry (ex,340; em,455)

TABLE 10.4
Examples of post-column derivatisation agents
(reproduced from Bristow (1976),
with permission)

Solute type	Reagent	Detection
Fatty acids	N,N'-dicyclohexyl-carbodiimide and hydroxylamine perchlorate; then ferric perchlorate	UV (530 nm)
Carbohydrates	Orcinol/sulphuric acid	UV (420 nm)
	Sulphuric acid	UV (300 nm)
Dicarboxylic acid	Chromic acid	UV
Amino acids	Ninhydrin	UV
	o-Phthalaldehyde	Fluorimetry (ex,340; em,455)
Kanamycins	o-Phthalaldehyde and fluorescamine	Fluorimetry (ex,320; em,450/ ex,375; em,480)
Phenols	Ce^{4+} sulphate	Fluorimetry

designs are popular, the specific choice being dependent upon the rate of the reaction, which ideally should be very fast (< 30 s). The first design consists of a simple coil which is supplied with both column eluent and reagent through a simple tee-junction and is ideal for reactions of less than 1 min. For reactions of up to 5 min packed beds of small glass beads (< 20 μm) are used (Snyder, 1976) and for longer reaction times a system of air segmented flow is required to minimise band broadening (Snyder, 1978; Scholten et al., 1981). It should be recognised that all these methods destroy the sample and therefore some form of eluent stream splitting is necessary if peak collection is required. Some examples of post-column derivatisation methods can be found in Table 10.4.

10.8. Column construction

All metals, including the 316 stainless steel used in most HPLC columns, eventually undergo chemical corrosion and halide ions in particular should be thoroughly removed when the HPLC system is not in use. It is good general practice to store all HPLC components in either pure solvents or in aqueous mixtures of organic solvents. It is a crucial requirement that all HPLC columns should have a smooth inner surface since irregularities can cause loss of performance and poor 'column-to-column' reproducibility. The most popular column dimensions are 150 mm × 4.6 mm I.D. for typical analytical applications, while much larger column diameters are available for preparative scale separations. More recently, for analytical applications there has been a great deal of interest in smaller diameter columns, e.g. microbore (0.5–1.0 mm I.D.) (Eckers et al., 1983); packed capillary (80–125 μm I.D.) and open tubular capillary (30–50 μm I.D.); these are discussed in greater detail in Section 10.9.

10.8.1. Frits

The inlet frit of an HPLC microparticle column is usually of 2 μm porosity and is the most frequent point for the collection of microparticulate matter, which eventually results in increased back pressures.

Since frits are relatively cheap the best remedy is to change the old frit for a new one. Alternatively, frits may be cleaned using either ultrasonication or treatment with strong acids. Most manufacturers supply columns with identical frits at each end and thus, although not usually recommended, column inversion can be a method whereby the top frit is flushed out. This method works well providing that the column is well packed and that no void volume exists at the original top end.

10.9. Microcolumns in HPLC

One of the most rapidly progressing areas in HPLC is that of microcolumn technology and advances in the not too distant future will probably result in microcolumn HPLC technology becoming more generally available. Research in microcolumn technology was initiated in the late 1960s by Horvath and Lipsky (1966) using columns which were 1 m in length and just 0.5 mm in diameter, and which had a packing of pellicular material. Subsequent developments resulted in these being superceded by columns with an internal diameter of 4–5 mm and a length of 15–30 cm. It is recognised that microcolumns possess intrinsic advantages over conventional columns and recent work by a number of different groups has led to the further development of microcolumns which may be broadly divided into three different categories: microbore (or small bore) packed columns, open tubular capillary columns and packed capillary columns.

10.9.1. Microbore packed columns

These columns are similar to normal HPLC columns except that they have an internal diameter of around 1 mm. The packing materials consist of small diameter particles (3–30 μm) packed into stainless steel columns by normal slurry packing techniques. An examination of the effect of internal diameter on column efficiency using 1 m columns packed with a silica gel stationary phase demonstrated that the lowest plate height values (H) were obtained with an internal diameter of approximately 1.02 mm (Scott and Kucera, 1979; McGuffin and Novotny, 1983). The effect of the particle sizes were also examined and the results suggested that a 20 μm packing produced the optimum reduced plate height curve, although both 5 μm and 10 μm packings gave higher column efficiencies. It should be emphasised that the optimum values for both internal diameter and particle size may only be valid for the specific stationary phase and extra-column dead volume of their equipment. It should also be noted that the sample capacity of microbore columns is much reduced compared with conventional columns because of their reduced cross-sectional areas.

10.9.2. Open tubular capillary columns

Open tubular capillary (OTC) columns are analogues of the capillary columns used in gas chromatography and usually consist of glass columns with very low internal diameters (50 μm or less) with either a bonded organic layer, an adsorbent layer or a mechanically deposited liquid layer uniformly distributed over the column wall. In theory, for the performance of OTC columns to match that of conventional packed columns, the internal diameter must be in the range of 10–30 μm.

The peaks eluted from OTC columns are of very low volume and this can create technical difficulties as the volumes of the flow cells in conventional detectors (5–10 μl) are much greater than the eluted peak volumes. If such detectors were used with OTC columns band dispersion could result, thereby negating the inherent advantages of these columns. Therefore, the flow cell volume should be in the range 0.01–1 μl. Similarly, there is a requirement in capillary systems for both a minimal dead volume and a reduction in the injection volume necessitating the development of specialised pre-concentration and sampling techniques. One further restriction of OTC columns is their small internal diameter which can result in sample overloading; only quantities of less than 10 ng should be used.

10.9.3. Packed capillary columns

Packed capillary (PC) columns were originally developed by Novotny and coworkers and may be regarded as a hybrid between packed microbore columns and OTC columns. PC columns have a larger internal diameter than OTC columns and contain a packing material which is evenly distributed within the column with a number of particles actually embedded in the column wall. The columns are constructed by packing alumina or silica gel (10–100 μm diameter particles) into a thick-walled glass column which is then drawn into a capillary. These columns show increased stability compared with OTC columns and column plate numbers in the region of 100,000 have been reported which indicates their potential for the resolution of complex samples.

10.9.4. General considerations

The foregoing suggests certain advantages of microcolumns over their conventional counterparts, but one aspect not discussed is the advantages which microcolumns may have in high speed chromatography. As microcolumns have only a small internal diameter, relatively low flow rates generate a high linear mobile phase velocity which, together with the high efficiency of microcolumns, can provide very rapid separations. Unfortunately, with such rapid separations, in addition to the requirement for a very low flow cell volume in the detector, the response time of the equipment must also be less than that found with most conventional detectors. However, under normal operating conditions microcolumns allow very efficient separations with very low flow rates and it has been demonstrated (Knox and Gilbert, 1979) that the speed of a separation using capillary columns will exceed that of conventional HPLC columns when the number of theoretical plates exceeds 20,000. Microcolumns can thus provide reductions in solvent consumption of between 90–99% with concomitant economic and environmental advantages. An additional advantage of microcolumns is their potential for high mass sensitivity which results from the peak volume being much reduced so that the mean relative solute concentration in the peak is greatly increased thus allowing the potential sensitivity of detection to be increased.

   In conclusion, the higher column efficiencies available with microcolumn HPLC could prove to be an absolute requirement for the analysis of complex biological samples. Technical advances in the next few years will presumably allow the introduction of the first commercially available microcolumn HPLC system.

10.10. Column packings

It has often been said that the heart of a good HPLC system is the column and clearly, irrespective of the excellence of the remainder of the system, if the column is inefficient, a good separation will not be achieved. In this section the influence of the column packing, including particle size and shape and the column porosity, is considered.

10.10.1. Particle size

Over the last decade much effort has been devoted towards increasing the efficiency of chromatographic separations to allow increased resolution in a reduced time. The essence of a chromatographic separation is to reduce band broadening of eluted peaks (as discussed in Chapter 2). The five major causes of band broadening are: (a) stagnant mobile phase mass transfer, (b) eddy diffusion, (c) longitudinal diffusion, (d) stationary phase mass transfer, and (e) mobile phase mass transfer. Of these parameters the stagnant mobile phase mass transfer is the most accessible to manipulation by variation in the particle size. Early packings consisted of large silica particles (30 μm diameter and above) which were totally porous and resulted in the formation of inaccessible deep pools of stagnant mobile phase. This problem was largely overcome by the development of pellicular packings (Horvath et al., 1967) which consisted of a solid core (usually glass) and a thin (1–2 μm) porous outer layer of stationary phase. These pellicular packings were therefore superficially porous and possessed only a shallow pool of stagnant mobile phase. However, the major disadvantage of pellicular columns was their low sample loading capacity due to the comparatively small amount of stationary phase which coated the particles. Clearly, the limits of detection of a compound in a complex mixture were considerably reduced as only a relatively small amount of sample could be loaded. More recently, totally porous microparticulate packings (3–10 μm particle diameter) have been developed which virtually eliminate stagnant mobile phase mass transfer problems and now pellicular packings are seldom used except in early method development. At present most analyses are carried out on columns which are packed with 10 μm or 5 μm particles with a recent tendency towards the latter. Theoretically, a reduction in particle size will substantially reduce column plate heights and increase column resolving power and it has been predicted that a particle diameter of about 2 μm will be optimum for many separations. Recently, columns with 3 μm particles have become commercially available and these show increased performance in terms of plate number per column metre and provide increased separation speeds (Fig. 10.2). However, a comparison of a 15 cm long, 5 μm particle size reversed phase column with a 7.5 cm long, 3 μm particle size column concluded that by increasing the mobile phase flow rate, equivalent resolution times could be achieved with the 5 μm particle size column with equivalent efficiency to the 3 μm particle size column although the pressure drop across the column was greater with the 5 μm particle size column (Cooke et al., 1982). In this instance the gain in resolution time was achieved at the expense of plate number, with the 5 μm particle size column having a greater plate number under optimal conditions. Thus, unless the extra efficiency of a 5 μm particle size column is required, the 3 μm particle size column may provide the shorter resolution time with a consequent reduction in consumption of mobile phases.

Four major problems are posed by a reduction in particle size below 5 μm diameter: (a) the very small volume in which peaks elute from the column, (b) viscous heating of the mobile phase with a consequent generation of thermal gradients within the column, (c) reduction in sample loading and (d) clogging of columns by particulates. As major advances have recently occurred in the development of very small particle columns each of these problems, with potential solutions, will now be considered.

The very small volume in which the solutes elute requires that the extra-column volumes are reduced to prevent extra-column dilution and band broadening. This means that the volume of the flow cell should be less than that used with conventional 5 μm particle size columns, thus causing a loss in mass detectability. The recent development of wide-bore 3 μm particle size columns (Zorbax Gold Series),

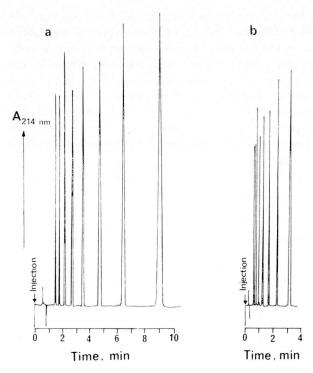

Fig. 10.2. Separation of 1-bromoalkanes (C_2–C_9) on 5 μm (150 mm × 4.6 mm I.D.) (a) and 3 μm (75 mm × 4.6 mm I.D.) (b) ODS columns. Chromatographic conditions: mobile phase, acetonitrile–water (80:20); flow rates, 1.4 ml/min (5 μm) and 2.0 ml/min (3 μm); temperature, 30°C; detection, UV at 214 nm. The same valve and sample loop were used to inject 5 μl of the same sample onto both columns. Reproduced from Cooke et al. (1982), with permission.

which have dimensions of 8 cm × 0.62 cm I.D., has been reported to reduce problems associated with narrow eluting bands and it may be calculated that providing the column dead volume is the same as with a 5 μm particle size column, the band volume will be similar. Thus a good performance may be obtained with conventional flow cells (8 μl) providing that the dead volumes between the injection valve and detector are minimised.

The generation of thermal gradients at high flow rates can also cause band broadening and remains a major problem with 3 μm particle size columns. One reported solution is to maintain the injection valve and the column at different temperatures; however, as the specific temperature difference will depend on the mobile phase flow rate, the system is not easily established. It has been reported that wide-bore (8×0.62 cm), 3 μm particle size columns which are air thermostatted experience only very small effects due to thermal gradients at flow rates of less than 10 ml/min. In general, however, with 3 μm particle size columns of conventional geometry, thermal gradients are a major problem.

The number of column plates (N) can be greatly reduced even under conditions where k' remains unaffected. A recent theoretical treatment (Poppe and Kraak, 1983) has concluded that the sample mass M_{max}^{10} for a 10% increase in band width (and a 20% decrease in N) is predictable as:

$$M_{max}^{10} = 2[(1 + k')/k']^2 W_s/N$$

where k' is the capacity factor for a small sample mass, N is the plate number and W_s is the weight of a sample taken up by a column at saturation. Thus with smaller particle columns, where N will be larger, it is clear that M_{max}^{10} will be smaller. The sample mass which can be loaded onto a 3 μm particle size column is therefore less than that of a 5 μm particle size column. A 8×0.62 cm, wide-bore column has a volume 1.8 times greater than that of a 0.46 cm diameter column of the same length, resulting in more stationary phase, a larger W_s component, and thus improved M_{max}^{10}. It should also be recognised that the diminished mass loadability of 3 μm particle size columns can cause problems with detection limits.

A problem common to all microparticulate columns is that of plugging by particulates and the precautionary procedures taken with 5 μm particle size columns become even more important with 3 μm particle size columns. Thus, the mobile phase and samples should be filtered (0.32–0.5 μm pore size), filters should also be present in-line between the pump and injector and a guard column should be introduced before the analytical column. Providing these precautions are followed, 3 μm particle size columns should cause little more trouble than 5 μm particle size columns.

10.10.2. Other aspects of particle size and shape

For columns packed by conventional techniques it is generally found that for particles of the same size approximately equivalent plate numbers are obtained with spherical and irregular particles but that the permeability of the spherical particle columns is nearly twice that of the irregular particle columns. However, by using careful packing techniques equivalent permeabilities have been reported for both irregular and spherical particle columns (Unger and Messer, 1978). No further consideration will be given to this aspect of column technology as nearly all commercially available column packings are spherical.

The problem of particle size distribution in a commercially supplied column packing material can pose problems to the chromatographer who packs his own columns. A large particle size distribution has no influence on column efficiency if the flow rate of the mobile phase through the column is maintained at optimum values. At higher flow rates there is a small negative effect on column efficiency, while at all eluting speeds a wide particle size distribution increases the column back pressure and the separation impedance. Thus, for a maximally effective chromatographic system the particle size distribution should be minimised.

10.10.3. Pore size

Size exclusion effects can influence the measurement of chromatographic parameters in all types of liquid chromatography. Size exclusion effects occur because of the existence of pores in the stationary phase which are filled with mobile phase and are too small for large sample molecules to enter. The efficiency of the column is clearly influenced by these effects as the rate of equilibration between the mobile and stationary phases depends on the rate of diffusion of the sample molecules into the pores. As the length of the alkyl chain attached to the particles constituting the stationary phase is increased, the rate of diffusivity into the pores is decreased due to restricted access. It is thus clear that an optimum situation must be achieved whereby the access of the sample molecules to the stationary phase is maximised without the appearance of size exclusion effects. Problems with size exclusion effects become particularly significant with protein and nucleic acid macromolecules, which have a tendency to aggregate, thereby causing not only size exclusion effects but also plugging of columns. In extreme cases, stationary phases with pore diameters of at least 50 nm are required.

10.11. Home-made column packings

10.11.1. Chemically bonded polymer phases

Commercially successful methods for the production of chemically bonded polymer phases suitable for HPLC have not been widely advertised and can only be referred to in general terms. Nevertheless, there are two principle methods for their production. Firstly, modification of a pre-formed support is possible providing that suitable sites are available. Alternatively, a method which has proved popular with polystyrene-type supports is to prepare the required support by co-polymerisation of appropriate monomer units.

10.11.2. Preparation of chemically bonded phases

Silica is the most popular support for adsorption chromatography because it is not only porous but also sufficiently rigid to withstand high pressures. These two features are essential for a good chromatographic support. For these reasons silica has been used as the support for chemically bonded phases although recent advances in organic hydrocarbon polymer technology has also provided chemically bonded phases to rival those based on silica. The wide pore size required for large molecules has meant that column pressures of greater than 1000 psi are rarely used and consequently semi-rigid polymers can be used.

10.11.3. Chemically bonded silica phases

Various forms of amorphous silica exist but all may be considered to have the general formula $Si(OH)_4$. Preparation of bonded phase

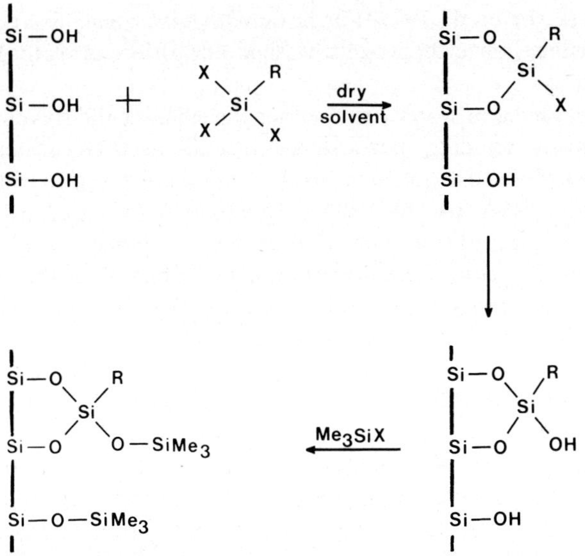

Fig. 10.3. Steps in the chemical derivatisation of silica. X = Cl or OR′ and R is one of the commonly used bonded phase such as C_{18}, C_8, C_2, cyano, amino, phenyl, etc.

materials relies on the exploitation of the surface hydroxyl groups of silica of which there are typically 5 per 10 nm^2 of surface. As the majority of the hydroxyl groups are sterically inaccessible to chemical reagents, bonded phase loadings are usually less than 1 mmol/g, representing 25% of surface sites. Fortunately, steric hindrance prevents unreacted sites from interfering in subsequent chromatography. This risk may be minimised by a final 'capping' reaction with a small reactive molecule.

A number of methods are available for the chemical modification of silica (Brust et al., 1973; Hastings et al., 1971) but the most popular is shown schematically in Fig. 10.3.

In practice, dry silica is refluxed for several hours with a solution of the organosilane ($RSiX_3$) in toluene or carbon tetrachloride under a dry inert gas atmosphere (Little et al., 1979a). The silica is recovered by filtration, washed (when X = Cl it is hydrolysed with water) and then capped to mask all unreacted silanol groups with trimethylchlorosilane (Little et al., 1979b) or hexamethyldisilazane. The chemically bonded silica, once thoroughly washed and dried, is ready for chromatography.

The presence of water (from either the silica or the solvent) during the chemical bonding, particularly with di- or trichlorosilanes, can cause hydrolysis and polymerisation of the silane on the surface of the silica, thus giving rise to a thicker coating as indicated by a higher carbon to silica ratio in elemental analysis. A higher carbon loading can in some circumstances be advantageous but often the coating is irregular and results in irreproducible resolution. This problem can be overcome using triethoxysilanes, where $X = OR'$.

Alternative bonded phases may be generated either by chemically modifying the attached group or by modifying the silane reagent prior to attachment to the bonded phase (Wheals, 1975; Chang et al., 1976).

Accurate quantitative analyses of bonded phases are difficult but the most successful has been provided by C, H and N analysis. As the carbon loading can have a major influence on chromatographic retention it is important that some indication of this parameter is provided and, indeed, most commercial suppliers of columns quote the carbon loading. Similarly, as the number of free hydroxyl groups can influence retention, especially in reversed phase chromatography, it is important to know whether or not a column has been end-capped.

10.12. Column packing

The cost of commercially pre-packed columns means that major financial savings can be achieved by packing columns 'in-house'. However, it should be realised that a poorly packed column can compact during use, causing the formation of column voids and impairing column performance. Corrective procedures can be carried out, such as filling the void with inert glass beads or a suitable pellicular material (pellicular materials are easier to dry pack provided that their diameter is greater than 20 μm). These procedures may also be used to remove deteriorated column packings (observed by discolouration). In general, slurry packing allows columns to be packed efficiently although, since HPLC pumps do not have the capability of generating the high impact velocities required, a high capacity, pneumatic amplifier pump is required (Majors, 1972).

Most commercial packing materials contain fines which need to be removed before packing. To achieve removal the support should be suspended in an organic solvent, allowed to settle for 15 min and then the supernatant should be decanted. This procedure should be repeated until the supernatant is clear and free of fines. Alternatively, the material can be washed on a filter paper which allows the fines to pass through but which retains the packing material (Whatman 3MM). The packing material can then be dried and suspended in the packing solvent. This solvent is usually of high density to prevent settling of the particles under gravity (e.g. a halogenated hydrocarbon). The suspension should be sonicated for five minutes to limit flocculation of the slurry and then poured into the packing reservoir prior to pumping into the HPLC column. About 3000 psi is recommended for packing and afterwards care must be taken to gradually reduce the pressure. The column can then be fitted with end-pieces, flushed with an appropriate solvent and tested for performance. For a review of alternative methods of column packing the reader is referred to Kaminski et al. (1982).

10.13. Column assessment

Most manufacturers can be relied upon to supply high quality columns with reproducible performances (Atwood and Goldstein, 1980) and usually a column is supplied together with a chromatogram of a test mixture under specified conditions. The column performance can be calculated from this test and should be equal to or above a guaranteed minimum. The chromatographer can check these specifications on receiving a column by running an identical sample to that specified. An inferior performance would indicate either that the column had been damaged in transit or that the extra-column effects of the user's equipment exceed those of the manufacturer's. It is also useful to run the test sample from time to time to monitor the column performance and for evidence of deterioration.

Column performance may be assessed using a number of criteria, the most useful being:
(1) The number of theoretical plates (N). This is a measure of the efficiency of the column in carrying out a particular separation. It is important to realise that this factor is dependent upon the test system

(i.e. flow rate, viscosity, sample, etc.) when comparing columns. However, the plate number alone is insufficient to specify the overall performance.

(2) Peak symmetry. This factor is important to ensure no loss of resolution. The peak asymmetry factor (A_s) is given by $A_s = b/a$ where b and a are the two peak half widths (b to the leading edge and a to the trailing edge) at 10% of peak height. The asymmetry factor may be reduced by raising the temperature or altering the mobile phase velocity.

(3) Column selectivity. This may be determined for a given pair of solutes and provides an indication of column-to-column reproducibility.

10.14. Column protection

The need to protect the chromatographic column from particulate matter by inclusion of in-line filters has already been described (Section 10.8.1) but there is an equivalent need to protect the column from some soluble components. Usually these materials would be removed during the sample preparation stage; however, as an added precaution a guard column can be included in the system (Lundanes et al., 1983). Guard columns are smaller than the main column. They are usually packed with a material similar to that in the main column and they effectively remove compounds with very high k' values which might otherwise clog the main column. Guard columns can be replaced frequently due to their relatively low cost and thereby prolong the lifetime of the main column. It should be noted that guard columns can cause a deterioration of column performance by contributing to extra-column band broadening effects, which should obviously be minimised.

The solvent used in the mobile phases can cause a progressive loss of column performance by slowly dissolving the column support material. This is a common problem with silica-based stationary phases operating at pH values above 8. Dissolution of the support gives rise to the appearance of column voids and results in loss of performance. This problem can be overcome by presaturation of the stationary phase by positioning a small column containing the same stationary phase as the main column upstream from the point of sample injection.

10.14.1. Column storage

Columns should usually be stored in a pure solvent with the ends of the column tightly capped. If ion-pair reagents or buffers have been used these should be thoroughly washed from the column with at least 50 column volumes to prevent the possibility of corrosion.

10.15. Tubing

The volume of the interconnecting tubing and fittings contributes to extra-column band broadening and therefore should be minimised. Generally, 1/16-inch steel capillary tubing is used which has an internal diameter of 0.5 mm, although an improved performance can be achieved with tubing with an internal diameter of 0.3 mm or 0.2 mm. Care should be taken in the use of tubing connectors such as the Swagelok system since each time the connection is undone it requires a little more force to subsequently re-seal. Continual opening and re-tightening or over-tightening in the first instance can seriously impair the flow of solvent, creating high back pressures. Where a seal is difficult to achieve PTFE tape may be wrapped around Swagelok fittings. It is also important not to create voids within these connectors as these can significantly increase extra-column band broadening. The main culprit for causing such voids is poor tubing ends and these should always be cut cleanly and perpendicularly and then smoothed off with a fine corundum stone. The tubing should then be ultrasonicated, flushed out and dried to remove any metal fines.

10.16. Detection

The various types of detector are described in the instrumentation section (see Chapter 3) and the specific choice of detector will depend on a number of criteria which are described in that section. However, a number of general points should be mentioned which can affect overall performances:

(a) Peak broadening effects can arise from the inherent detector cell volume and connecting tubing. Thus, detector cell volume should be minimised to a level which retains maximal sensitivity.

(*b*) Response time should be sufficiently fast to allow the detector output to reflect the true chromatographic eluate.

(*c*) Compound detectability. It is important to establish that the compound of interest can be detected by the detector. This will depend on the physicochemical characteristics of the compound.

(*d*) Background noise should be minimised to improve detection sensitivity (particularly for trace analysis). Mains voltage peak smoothing devices are recommended for electrically 'noisy' environments.

10.17. Peak identification

Retention time and co-chromatography with reference compounds are a very useful guide to peak identification but require further confirmation and a number of methods have been employed for this purpose including:

(*a*) Chemical or physical characterisation (e.g. mass spectrometry).

(*b*) UV absorbance ratios.

(*c*) Stopped flow UV/fluorescence scanning.

(*d*) Enzymatic peak shift.

(*e*) Derivatisation.

These various methods can be used either separately or in combination (Krstulovic et al., 1978).

10.18. Data handling

There are two methods for the quantitation of material in chromatographic peaks:

(1) Direct use of the recorder trace to measure peak height (this is approximately proportional to peak area) or peak area (particularly for gradient work where a non-stable baseline may be experienced). The former method is simple and quick but obviously has in-built inaccuracies due to asymmetric peaks.

(2) Electronic integration. Obviously this method is more expensive but is much more convenient and more accurate. Unfortunately the detector response factors vary for different compounds and this must be taken into account. Derivatisation is a good way of equalising response factors.

CHAPTER 11

Applications of HPLC

Nucleosides and nucleotides

11.1.1. Introduction

Liquid chromatography found one of its earliest applications in the separation of nucleosides and in part was responsible for the rapid development in the understanding of nucleic acid biochemistry. Subsequently, chromatography has found a wide application in the determination of nucleic acid composition, in elucidating nucleic acid metabolism and also in quantitating nucleotide pools. The earliest report of column liquid chromatographic separation of nucleic acid components can be found in Cohn (1949). Since then, liquid chromatography has been the most widely employed method for the separation of these components. The standard method for separating bases, nucleosides and nucleotides was by ion-exchange chromatography on simple polystyrene-type ion-exchange resins.

In 1967, Horvath developed pellicular ion-exchange stationary phases which were both stable and relatively efficient and enabled the application of higher pressures, thereby facilitating the rapid separation of the mono-, di- and triphosphates of cytidine, uridine, adenosine, thymidine and guanosine. The subsequent development of stationary phases based on porous silica has led to a variety of chromatographic modes which are able to rapidly separate nucleosides and their derivatives.

The HPLC analysis of nucleic acid components can be divided into sections according to the chemical nature of the components. For example, nucleic acids are composed of equal proportions of an organic base, a carbohydrate and a phosphate residue, covalently linked into long polymeric chains and it is this variety of characteristics which determine the most appropriate chromatographic mode. In this chapter we will consider the components in the following sections:

nucleoside bases,

nucleosides,

nucleotides, and

polynucleotides (RNA and DNA).

Natural nucleoside bases can be one of five heterocyclic aromatic compounds and include adenine, guanine (the purines), cytosine, thymine and uracil (the pyrimidines). Nucleosides are composed of a base which is attached to a carbohydrate moiety through one of the aromatic ring nitrogen atoms. The specific carbohydrate distinguishes RNA (ribose) from DNA (deoxyribose). Nucleotides are phosphorylated derivatives of nucleosides and contain one, two, three or (less commonly) four or more phosphate residues either at the 5' position or at the 3' position (Fig. 11.1.1).

The three HPLC modes commonly used for the separation of these compounds include ion-exchange, reversed phase and reversed phase ion-pair. The pK_a values of the bases and nucleosides play a major role in determining their retention times in all chromatographic modes and a list is presented in Table 11.1.1 for reference (pK_{ab} and pK_{aa} refer to the first gain and to the first loss of a proton).

A number of chemically modified nucleoside derivatives have

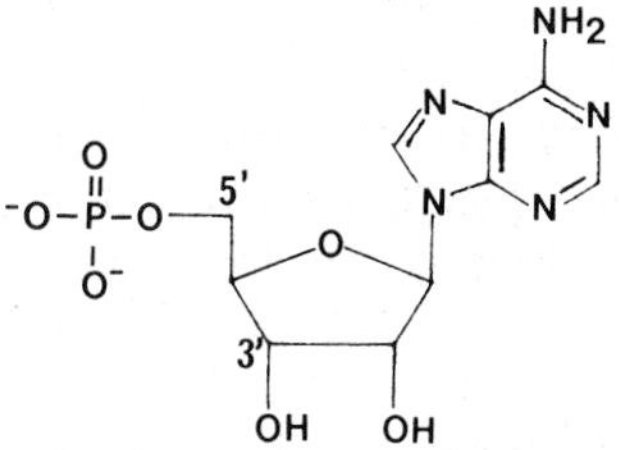

Fig. 11.1.1. Structure of adenosine ribonucleoside 5'-monophosphate.

TABLE 11.1.1
pK_a values of bases and nucleosides

	pK_{ab}	pK_{aa}
Bases		
Adenine	4.15	9.8
Guanine	3.20	9.6
Hypoxanthine	2.00	8.9
Xanthine	0.80	7.5
Cytosine	4.45	12.2
Uracil	−3.40	9.5
Thymine	–	9.9
Nucleosides		
Adenosine	3.50	12.5
Guanosine	1.60	9.2
Inosine	1.20	8.8
Xanthosine	< 2.50	5.7
Cytidine	4.15	12.5
Uridine	–	9.2
Thymidine	–	9.8

important biological functions and several of these compounds are used for the treatment of viral infections and cancer; these will also be discussed in this chapter.

11.1.2. Sample preparation

The extraction of nucleosides and nucleotides from biological sources is conventionally carried out by homogenisation in ice-cold perchloric acid (or less often trichloroacetic acid) (Maybaum et al., 1980). This has the benefit of precipitating the protein from the sample, leaving the nucleotides in the supernatant which then can be removed after centrifugation. The supernatant can be neutralised with an alkaline solution such as an aqueous hydroxide or an amine–freon solution. More recent techniques include the removal of protein by ultrafiltration which eliminates the potential for chemical modification. Alternatively, deproteinisation can be carried out by treatment with ice-cold acetonitrile.

It is possible to separate deoxyribonucleosides from ribonucleosides by utilising the difference in the chemical properties of the carbohydrate portion of the molecule. Periodate solution will specifically react with the *cis*-diol function in ribonucleosides and ensure their effective removal (Garrett and Santi, 1979). Alternatively, ribonucleosides can be specifically bound to a boronate gel column such as Affi-Gel 601 at basic pH and later eluted with 0.1 M formic acid (Colonna et al., 1983).

11.1.3. Characterisation of eluent peaks

The classical methods for the identification of nucleoside eluent peaks is to use either retention time or 'spiking' (co-chromatography) with reference compounds. In addition, chemical treatment with periodate can be used to identify ribonucleosides by observing the lost peaks from a second chromatogram (Hartwick et al., 1979b). Similarly, enzymatic modification can be useful in a number of specific cases (Table 11.1.2).

The loss or decrease in area of the substrate peak or increase in area of the product peak aids compound identification.

The UV absorbance ratio (A_{254}/A_{280}) provides an alternative means for compound identification. Some characteristic ratios are presented in Table 11.1.3.

The nucleoside derivatives obtained from biological sources are often at very low levels and this can be a serious problem not only for the sensitivity of UV detection but also for identification purposes.

TABLE 11.1.2
Uses of enzymatic modification

Enzyme	Substrate	Product
5'-Nucleotidase	Nucleoside monophosphate	Nucleoside
Alkaline and acid phosphatase	Nucleotide	Nucleoside
Purine nucleoside phosphorylase	Nucleoside + P_i	Base
Xanthine oxidase	Hypoxanthine	Xanthine
	Xanthine	Uric acid
Adenosine deaminase	Adenosine	Inosine
Guanase	Guanine	Xanthine
Adenosine kinase	Adenosine	Adenosine monophosphate

Reproduced from Zakaria and Brown (1981), with permission.

TABLE 11.1.3
UV absorbance ratios

Compound	A_{254}/A_{280}
Adenosine	4.71
1-Methyladenosine	4.33
Cytidine	0.92
3-Methylcytidine	0.33
5-Methylcytidine	0.51
Guanosine	1.60
1-Methylguanosine	1.84
2-Methylguanosine	1.80
7-Methylguanosine	1.36
Inosine	5.89
1-Methylinosine	3.66
Pseudouridine	2.17
4-Thiouridine	1.73
Uridine	2.55

Reproduced from Gehrke et al.
(1980), with permission.

Chloroacetaldehyde undergoes a highly specific reaction with adenine and related compounds to produce $1,N^6$-ethenoadenine, a highly fluorescent derivative with an emission maximum at 410 nm (Leonard and Tolman, 1975). This method allows the specific detection of 1 pmol of adenine and related nucleosides and nucleotides (Preston, 1983).

11.1.4. Nucleoside bases

Nucleic acid bases are often analysed in the presence of the corresponding nucleosides since the preferred chromatographic modes are suitable for the simultaneous separation of both classes of compounds. Originally the polar character of the nucleoside bases was exploited using ion-exchange HPLC (Floridi et al., 1977); however, reversed phase techniques are now more commonly employed.

In two of the most thorough investigations into chromatographic techniques for the separation of nucleoside bases, nucleosides and other UV-absorbing compounds, the retention data for 86 compounds on a reversed phase column has been reported both qualitatively

(Hartwick et al., 1979a) and quantitatively (Assenza and Brown, 1983). Excellent separations of up to 28 compounds were obtained on a single ODS column using an aqueous phosphate buffer–methanol gradient. A relationship was shown between chemical structure and retention time in the purine and pyrimidine series, allowing retention time to be predicted from the chemical structure.

The accurate determination of the concentrations of purine and pyrimidine metabolites in a variety of biological fluids allows the study of a number of disease states and may suggest therapeutic approaches. Using reversed phase chromatography it is possible to separate nucleoside bases, nucleosides and nucleotides in a single chromatographic run (De Abreu et al., 1982). Samples were derived from physiological fluids such as urine, serum and plasma. Urine was used directly after filtration but serum and plasma samples were first deproteinised by precipitation with perchloric acid and then neutralised before injection. The HPLC separation was developed with test compounds on an ODS stationary phase utilising a step-wise ternary solvent gradient starting with 0.05 M phosphate buffer (pH 5.6) and ending with 0.05 M phosphate buffer (pH 5.6)–methanol–water (50 : 25 : 25). The application of this technique to samples of urine from a Lesch-Nyhan syndrome patient is shown in Fig. 11.1.2. Eluted components were monitored by a UV detector at 254 nm and 280 nm and permits detection of 5–10 pmol of each component. The metabolic fate of radiolabelled precursors can be monitored by using a radioactivity detector (Webster and Whaun, 1981).

A similar example in which 5-fluorouracil has been included demonstrates the wide applicability of reversed phase systems coupled with aqueous phosphate buffers for the separation of nucleoside bases (Miller et al., 1982). This study also includes a useful comparison of nine different analytical reversed phase columns in combination with different isocratic conditions and clearly demonstrates subtle differences between them. The study concluded that the best stationary phase to maximise the resolution of the test compounds was Spherisorb ODS-2 in combination with ammonium phosphate (pH 3.5) as eluent buffer (Fig. 11.1.3).

Methanol or acetonitrile are occasionally used as organic modifiers in the reversed phase separations of nucleoside bases and have the

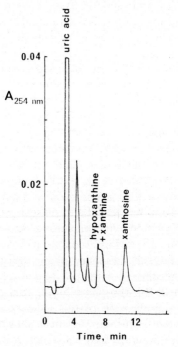

Fig. 11.1.2. Analysis of urine from a Lesch-Nyhan syndrome patient. Chromatographic conditions: column, Spherisorb 10-ODS (250×4.6 mm); mobile phase, ternary gradient from 0.05 M potassium phosphate, pH 5.6 to 0.05 M potassium phosphate, pH 5.6−methanol−water (50:25:25); flow rate, 1.5 ml/min; temperature, 40°C; detection, UV at 254 nm. Reproduced from De Abreu et al. (1982), with permission.

effect of reducing retention times and altering selectivities (Hartwick et al., 1979b). Acetate buffers have also been employed as an alternative to phosphate buffers but give relatively low column efficiencies, possibly because of the formation of non-polar complexes with the solute molecules. The pH of the eluent buffer has little effect on retention unless a change in the ionisation state of the nucleoside base occurs. This change can have a dramatic effect on retention time and allows fine control over selectivity.

The reversed phase mode is adequate for most separations involv-

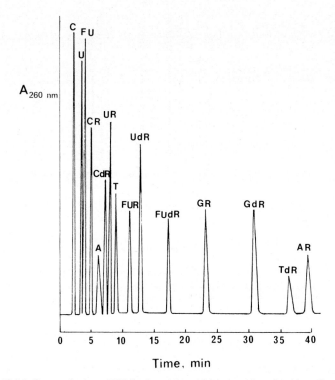

Fig. 11.1.3. Reversed phase HPLC of nucleic acid bases and nucleosides. Chromatographic conditions: column, Spherisorb ODS-2, 5 μm (250×4.6 mm); mobile phase, 0.05 M monobasic ammonium phosphate, pH 3.5; flow rate, 1.5 ml/min; temperature, ambient; detection, UV at 260 nm. Peaks: C, cytosine; U, uracil; FU, fluorouracil; CR, cytosine riboside; A, adenine; CdR, cytosine deoxyriboside; UR, uracil riboside; T, thymine; FUR, fluorouracil riboside; UdR, uracil deoxyriboside, FUdR, fluorouracil deoxyriboside; GR, guanine riboside; GdR, guanosine deoxyriboside; TdR, thymine deoxyriboside; AR, adenine riboside. Reproduced from Miller et al. (1982), with permission.

ing nucleoside bases although for some modified bases resolution from the solvent front can be a problem. This is simply overcome by employing the reversed phase ion-pair mode. Thus, using a mobile phase containing 5 mM heptanesulphonate in 2.5 mM potassium phosphate (pH 5.6), it was possible to resolve cytosine and 5-methyl

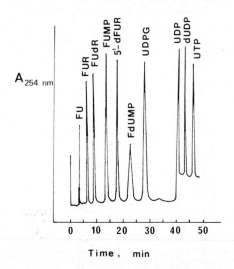

Fig. 11.1.4. Separation of uracil and 5-fluorouracil bases, nucleosides and nucleotides by reversed phase ion-pair HPLC. Chromatographic conditions: column, μBondapak C_{18} (300 × 4 mm); mobile phase, (from 0–30 min) 0.1 mM tetrabutylammonium hydrogen sulphate (C_{16}), 2.5 mM tetraethylammonium bromide (C_8) and 2% methanol in 2 mM sodium acetate, 1.5 mM phosphate buffer, pH 6.0 (Buffer A); (from 30–50 min) Buffer A + 30 mM phosphate; detection, UV at 254 nm. Peaks: FU, fluorouracil; FUR, fluorouracil riboside/ FUdR, fluorouracil deoxyriboside; FUMP, fluorouridine 5′-monophosphate; 5′dFUR, 5′-deoxyfluorouracil riboside; FdUMP, deoxyfluorouridine monophosphate; UDPG, uridine diphosphoglucose; UDP, uridine diphosphate; dUDP, deoxyuridine monophosphate; UTP, uridine triphosphate. Reproduced from Au et al. (1982), with permission.

cytosine (Erlich and Erlich, 1979). In an elegant study of the factors affecting reversed phase ion-pair HPLC, the separation of the bases, nucleosides and nucleotides of 5-fluorouracil and its analogue 5′-deoxy-5-fluorouridine was achieved in a two-step elution procedure (Fig. 11.1.4). The preferred mobile phase contained 0.1 mM tetrabutylammonium ions, 2.5 mM tetraethylammonium ions and 2% methanol in a 2 mM sodium acetate–1.5 mM phosphate buffer, pH 6.0.

The separation of nucleosides and their corresponding bases has also been achieved on unmodified silica. The retention times of at least fifty biologically important compounds, including nucleoside derivatives, have been reported (Ryba and Beranek, 1981). Several pairs of components which were not readily resolved using reversed phase chromatography were separated on silica by utilising a mixture of dichloromethane, methanol and aqueous buffer; the low solubility of nucleosides in such systems limited the application to very low sample sizes (5–10 μg). A similar study investigated the effects of pH and mobile phase content and concentration on retention (Brugman et al., 1982).

11.1.5. Nucleosides

Nucleosides play a vital role in many biological systems, not only as precursors to DNA and RNA but also in their own right, functioning as metabolic regulators. Consequently, the quantification and HPLC analysis of nucleosides in biological fluids can provide important information regarding their function. The analysis of the major and minor components of DNA and RNA is most conveniently carried out at the nucleoside level and will be considered as a separate section (Davis et al., 1979).

In the past HPLC separations of nucleosides have been carried out in a variety of modes, including anion-exchange (Floridi et al., 1977) and cation-exchange (Breter et al., 1977); however, since the introduction of stable packings reversed phase has become the preferred method. Typical chromatographic conditions for the separation of nucleosides include the use of dilute phosphate buffers with organic modifiers such as methanol or acetonitrile on ODS stationary phases. The effects of variations in these parameters is described below.

11.1.5.1. Effect of pH

Variation of pH in the range between 4.0 and 8.0 has little effect on the retention time of the common nucleosides. There is a slight trend towards longer retention times as the pH is increased within this range but more dramatic effects occur for those nucleosides with pK_a values in this region. For example, 3-methyl cytidine, 5-methyl cytidine, 1-methyl adenosine, adenosine and 7-methyl guanosine have pK_a values of 8.7, 4.3, 7.6, 3.5 and 7.1, respectively, and consequently their retention times change disproportionately as they gain or lose their charge. Thus, pH can be used to modify the selectivity of the chromatographic system.

11.1.5.2. Effect of organic modifier

Increasing the concentration of organic solvent in the mobile phase from 0 to 10% generally decreases the retention times of the common nucleosides in a logarithmic fashion. It has been possible to divide nucleosides into four different groups depending on the response of their selectivity factors to increasing methanol concentrations; within such groups there is little variation in selectivity (Gehrke et al., 1980).

11.1.5.3. Effect of temperature

The retention time of nucleosides decreases as a linear function of the increase in temperature in the range 25 to 55°C. The column efficiency significantly increases with increasing temperatures.

A useful technique worth noting here, but which is also valuable for analysis of nucleotides, is the use of boronic acid affinity columns for the separation of molecules such as ribonucleosides and ribonucleotides containing a 2′,3′-cis-diol from molecules not containing this functionality such as deoxyribonucleosides and 2′,3′-cyclic nucleoside monophosphates. At alkaline pH molecules containing the 2′,3′-cis-diol moiety will form a complex with the supported boronate groups (Hageman and Kuehn, 1977) and by lowering the pH the adsorbed fraction can be eluted and volatile buffers removed by lyophilisation (Buck et al., 1983). This technique has been adapted for use in HPLC by attaching a benzene boronic acid group via a spacer molecule to porous silica (Glad et al., 1980). The commercially available supports include Bio-Rad Affi-Gel 601.

The general principles which have been outlined above are applicable to all nucleoside separations but it is instructive to consider specific examples drawn from the various areas in which these techniques are valuable.

11.1.5.4. Natural nucleosides

11.1.5.4.1. In biological fluids and tissues. The value of HPLC in the analysis of nucleosides in biological tissues and fluids arises from the

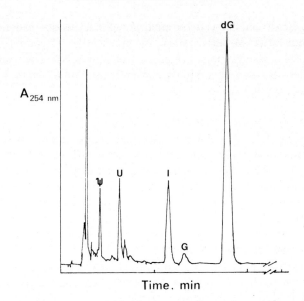

Fig. 11.1.5. HPLC profile of nucleosides in normal human serum. Chromatographic conditions: column, μBondapak C_{18} (300×4 mm); mobile phase, 0.01 M monobasic ammonium phosphate, pH 5.1 containing 6% methanol; flow rate, 1 ml/min; temperature, 36°C; detection, UV at 254 nm. Peaks: ψ, pseudouridine; U, uridine; I, inosine; G, guanosine; dG, deoxyguanosine. Reproduced from Colonna et al. (1983), with permission.

resolution and sensitivity afforded by the technique and the possibility of unambiguous identification of the compounds. After deproteinisation, chromatography has been carried out on a μBondapak C_{18} reversed phase column (Colonna et al., 1983) (Fig. 11.1.5).

Relatively few papers have reported the estimation of nucleosides in animal cells using HPLC. However, the metabolic state of cells can be determined in this way and this allows the progress of certain diseases to be monitored (Nissinen, 1980). A recent determination of nucleosides in cell extracts was carried out after initial acid precipitation, neutralisation and centrifugation followed by chromatography on a reversed phase C_{18} column using an aqueous sodium acetate–acetonitrile mobile phase (Harmenberg, 1983).

In the presence of relatively high nucleotide levels (e.g. as can be found in heart cells) the determination of nucleosides can best be carried out after an extra stage of purification (Henderson and Griffin, 1981) using anion-exchange chromatography. The separated nucleosides can then be analysed conventionally on a reversed phase C_{18} stationary phase.

11.1.5.4.2. As components of DNA.   Most DNA molecules, whether from prokaryotes or eukaryotes, contain some minor methylated components produced by post-replicational processing. 5-Methylcytosine and N^6-methyladenine often occur in prokaryotes while only the former is usually found in higher eukaryotes (Hall, 1971). The content of these minor bases varies between 0.05 mole percent and 7 mole percent depending on the organism. Methylation of DNA by various agents is suspected to be involved in tumor formation and consequently much research has centred on the determination of methylated nucleosides in DNA. Quantitation by paper chromatography and visualisation by UV absorption or radioactivity were early methods which have been superceded by cation-exchange HPLC (Lapeyre and Becker, 1979), reversed phase HPLC (Wakizaka et al., 1979) and reversed phase ion-pair HPLC.

A reversed phase procedure offering good sensitivity, selectivity and precision for the determination of all six deoxyribonucleosides with a mild sample preparation method has been reported (Kuo et al., 1980). The DNA samples were degraded to their component deoxyribonucleosides by complete digestion with the enzymes DNAase 1, nuclease P1 and bacterial alkaline phosphatase. Complete separation of the six deoxyribonucleosides was achieved in 70 min on a μBondapak C_{18} reversed phase support using dual wavelength UV detection.

An isocratic reversed phase system for the separation of authentic major and minor DNA constituents in 14 min has been reported (Assenza et al., 1983). The relatively short analysis time was attributed to the use of a trimethylsilyl reversed phase packing (Fig. 11.1.6). In the presence of 6-methyladenosine the time required for complete resolution was extended to 40 min.

The 5-methylcytidine content of DNA has recently been determined after complete acid hydrolysis to the constituent bases which

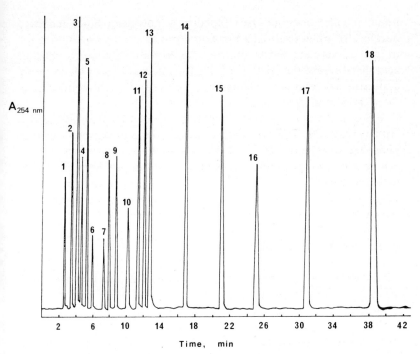

Fig. 11.1.6. Isocratic separation of major and minor ribonucleosides and deoxyribonucleosides. Peaks: 1, cytidine; 2, deoxycytidine; 3, inosine; 4, guanosine; 5, 5-methyldeoxycytidine; 6, deoxyinosine; 7, deoxyguanosine; 8, 7-methylguanosine; 9, thymidine; 10, adenosine; 11, 1-methylguanosine; 12, 7-methyldeoxyguanosine; 13, deoxyadenosine; 14, 1-methyldeoxyguanosine; 15, 2,2-dimethylguanosine; 16, 6-methyladenosine; 17, 2,2-dimethyldeoxyguanosine; 18, 6-methyldeoxyadenosine. Chromatographic conditions: column, Zorbax TMS (6 μm) (250×4.6 mm); mobile phase, 2% methanol in 4 mM di-ammonium hydrogen phosphate and 4 mM ammonium di-hydrogen phosphate, pH 4.0; flow rate, 2.0 ml/min; temperature, 35°C; detection, UV at 254 nm. Reproduced from Assenza et al. (1983), with permission.

were then separated by cation-exchange HPLC using a 0.02 M phosphate buffer at pH 2.3 (Diala and Hoffman, 1982).

11.1.5.4.3. As components of RNA.   Modified nucleic acid bases occur naturally as constituents of several cellular RNA species and in particular in tRNA, where they can comprise 25% of the total number of nucleosides. Their appearance in tRNA occurs by post-transcriptional modification but their function is not clear. Since modified nucleosides are not salvaged by natural processes, their quantification in body fluids can provide a method for monitoring abnormal nucleic acid metabolism; the analysis of modified nucleosides in blood serum and urine has been used as a method for diagnosis and monitoring of the treatment of cancer patients (Davis et al., 1977).

Release of the nucleoside components from tRNA can be carried out by enzymatic hydrolysis with nuclease P1 and bacterial alkaline phosphatase (Gehrke et al., 1982). Quantitative hydrolysis can be achieved in two hours and does not suffer from problems associated with the chemical lability of certain nucleosides (Randerath et al., 1972). A similar analysis of the nucleosides in *Salmonella typhimurium* tRNA hydrolysates has been reported (Buck et al., 1983).

11.1.5.5. Synthetic nucleosides

Modified nucleosides have found applications in the treatment of a variety of disease states; for example, 5-substituted pyrimidine nucleosides are used as antiviral agents and arabinocytidine is used in anticancer therapy. The quantitation of these compounds and their metabolites in biological fluids is essential for optimising dosing schedules. The separation of a mixture of ribosyl, deoxyribosyl- and arabinosyl purines without the need to use periodate or borate concentration procedures has been reported. The eluting peaks were identified using the peak shift method with the enzyme adenosine deaminase and monitoring the production of the appropriate hypoxanthine derivatives (Agarwal et al., 1982).

Conventional reversed phase HPLC is unable to separate ara-cytidine, ara-uridine, ribo-cytidine and ribo-uridine whereas a dual column procedure utilising an Ultrasphere ODS column in series with a Partisil PXS 10/25 SCX column gave excellent separation of these compounds (Sinkule and Evans, 1983). The accuracy of the method allowed comparative pharmacokinetic and pharmacodynamic studies to be carried out.

Other references to reversed phase analysis of synthetic modified nucleosides can be found in the literature (Delia et al., 1982; Robinson et al., 1985).

11.1.6. Nucleotides

Nucleotides play a central role in the regulation of metabolic pathways and their quantification in biological samples is of fundamental importance. There is an enormous variety of nucleotides (i.e. the four

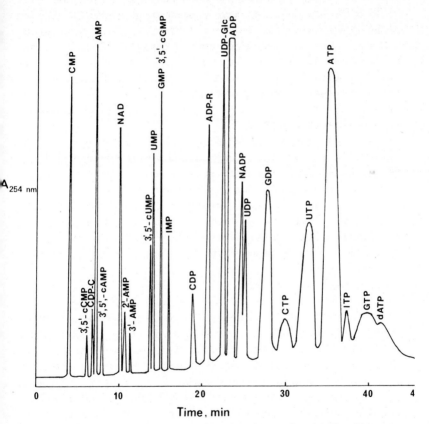

Fig. 11.1.7. HPLC anion-exchange separation of a mixture of nucleotides. Chromatographic conditions: column, μBondapak NH$_2$; mobile phase, linear gradient from 0.005 M potassium dihydrogen phosphate, pH 3.0 to 0.5 M potassium dihydrogen phosphate, pH 4.0 over 54 min; flow rate, 1.3 ml/min; detection, UV at 254 nm. Reproduced from Brown et al. (1982), with permission.

major nucleosides as either their monophosphates, diphosphates, tri-phosphates or cyclic phosphates) and their resolution, identification and quantitation is difficult since they are very similar in physicochemical terms. The dominant chemical influence comes from the presence of the phosphate groups which are ionised even at pH values near neutrality and has resulted in the majority of nucleotide separations being carried out by anion-exchange (Garrett and Santi, 1979). However, there has been a recent trend to use the reversed phase mode for the resolution of nucleotides due to the greater stability of this stationary phase and for the greater flexibility in the choice of mobile phase (Hodge and Rossomando, 1980). The specific choice of chromatographic mode for a given application remains controversial and there are papers extolling the virtues of both ion-ex-change (Pogolotti and Santi, 1982) and reversed phase (Hull-Ryde et al., 1983) chromatography.

The relative merits of ion-exchange and reversed phase modes are discussed in a study of the nucleotide pools from mammalian tissues (Brown et al., 1982). The tissue extracts were initially purified on a Cu^{2+}-loaded Chelex 100 resin to maximise both resolution and sensi-tivity during the main HPLC separations. It was concluded that the best routine method was on the anion-exchange column containing μBondapak NH_2 (Fig. 11.1.7), but where necessary this could be complemented with reversed phase or reversed phase ion-pair HPLC.

11.1.6.1. Ion-exchange HPLC

Typically, nucleotides have been analysed on strong anion-exchange stationary phases such as Partisil 10-SAX, using a mobile phase containing an acidic phosphate buffer. Thus, the complete resolution of the mono-, di- and triphosphates of adenosine, cytidine, guanosine, inosine, uridine and thymidine was achieved on Partisil 10-SAX using a mobile phase gradient from 0.007 M KH_2PO_4 (pH 4.0) to 0.25 M KH_2PO_4, 0.25 M KCl (pH 4.5) over a period of 80 min (Hartwick and Brown, 1975).

The selectivity of ion-exchange chromatography for nucleotide separations is significantly affected by the pH of the mobile phase and this can therefore be manipulated to achieve many separations.

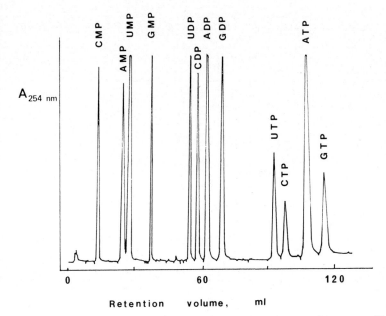

Fig. 11.1.8. HPLC profile of a standard mixture of 12 ribonucleotides, chromatographic conditions: column, Partisil 10-SAX (250×4.6 mm); mobile phase, Buffer A (7 mM sodium dihydrogen phosphate, pH 3.8), Buffer B (250 mM sodium dihydrogen phosphate, pH 4.5), Buffer A for 6 min, linear gradient for 30 min, buffer B for 10 min; flow rate, 3 ml/min; detection, UV at 254 nm. Reproduced from Pogolotti and Santi (1982), with permission.

Values between pH 3.5 and pH 5.5 appear to offer the best selectivity (Ericson et al., 1983). The ion-exchange mode has been used successfully for the separation of the major nucleotides from L1210 cells (Fig. 11.1.8). From this system the calculated chromatographic parameters for over 30 components were calculated (Pogolotti and Santi, 1982). For preparative ion-exchange separations of nucleotides a volatile eluent is desirable and the use of ammonium carbonate and ammonium acetate have been described (Linz, 1983).

11.1.6.2. *Reversed phase HPLC*

At the pH values commonly used in reversed phase HPLC the nucleotides are negatively charged and generally this mode is used for

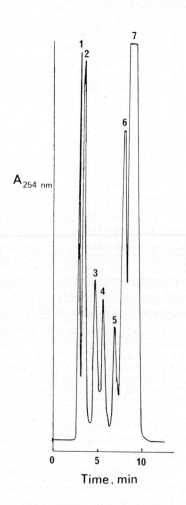

Fig. 11.1.9. Separation of a solution of adenosine and its phosphorylated derivatives. Chromatographic conditions: column, Ultrasphere 5 μm ODS (250×4.6 mm); mobile phase, 0.05 M ammonium dihydrogen phosphate, pH 5.0 in methanol–water (10:90); flow rate, 1.9 ml/min; detection, UV at 254 nm. Peaks: 1, 5'-ATP and 5'-ADP; 2, 5'-AMP; 3, 2'-AMP; 4, 2',3'-cAMP; 5, 3'-AMP; 6, 3',5'-cAMP; 7, adenosine. Reproduced from Ramos and Schoffstall (1983), with permission.

'group' separations in which there are distinct differences in their properties. Individual nucleotides are not retained on reversed phase stationary phases and using water as the mobile phase on an ODS stationary phase, 2'- and 3'-CMP, 2'- and 3'-UMP, 2'-GMP, UTP and UDP all eluted in the void volume with only cAMP and uridine being retained (Kessler, 1982). However, increasing the ionic strength by the addition of an acidic phosphate buffer, the individual nucleotides were retained and resolved. The addition of an organic modifier such as methanol or acetonitrile (5%) to a phosphate buffer (pH 5.0) facilitates the resolution of nucleotides while maintaining a very stable baseline (Debetto and Bianchi, 1983).

Reversed phase HPLC has been used successfully for the separation of the four monophosphates of adenosine, namely the 2'-, 3'-, 5'- and the 2',3'-cyclic phosphates (Fig. 11.1.9). The method is equally useful for resolution of the nucleotides of 2'-deoxyadenosine, 2'-deoxythymidine and 3'-deoxyguanosine (Ramos and Schoffstall, 1983). The added organic solvent concentration in the mobile phase strongly influenced the separations with higher concentrations reducing retention, presumably by competing for the sites on the stationary phase. The order of elution of the nucleotides generally correlated with the hydrophobicity of the compounds.

11.1.6.3. Reversed phase ion-pair HPLC

The reversed phase mode has obvious limitations for the separation of nucleotides because of the negative charge carried by most nucleotides at commonly used pH values which can result in both poor retention and lack of resolution. The addition of an ion-pairing reagent to the mobile phase may be used to overcome these problems. Thus, using a gradient mobile phase which contained tetra-n-butylammonium hydrogen sulphate as an ion-pairing reagent, twelve nucleotide phosphates were resolved (Hoffman and Liao, 1977).

A comparison of ion-exchange, reversed phase and reversed phase ion-pair chromatography in the separation of nucleotides from bacterial cells concluded that the latter method was preferable for analytical separations (Fig. 11.1.10). The resolution obtained allowed

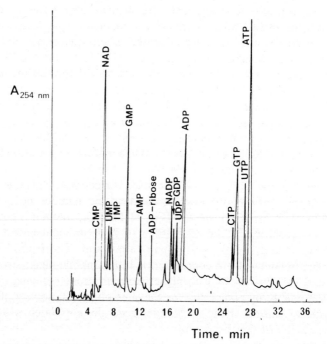

Fig. 11.1.10. Separation of nucleotides by reversed phase ion-pair chromatography. Chromatographic conditions: column, Ultrasphere 5 μm ODS (250×4.6 mm); mobile phase, Buffer A (5 mM tetrabutylammonium phosphate, 4% acetonitrile and 50 mM potassium dihydrogen phosphate, pH 6.0), Buffer B (acetonitrile), a gradient from 0–40% buffer B was used over 60 min; flow rate, 1.5 ml/min; detection, UV at 254 nm. Reproduced from Payne and Ames (1982), with permission.

detailed analysis of the effect of growth conditions on the intracellular nucleotide levels.

Optimal mobile phase conditions for the separation of nucleotides using reversed phase ion-pair chromatography have been found to be a phosphate buffer which contains tetrabutylammonium phosphate (TBAP) as the counter-ion (Darwish and Prichard, 1981). Phosphate concentrations between 50 mM and 120 mM gave the best resolution, whilst the addition of TBAP at concentrations between 0.5 mM and 2.0 mM significantly increased resolution and retention time. Concentrations greater than 2.0 mM increased retention without improving resolution. The addition of an organic solvent (notably acetonitrile) can be used to reduce retention whilst having little effect on resolution. Variations of the pH value of the mobile phase can be used to modify selectivity between closely related pairs of nucleotides.

11.1.6.4. Normal phase HPLC

Natural nucleotides are too polar to allow chromatography in the normal phase mode but chemically modified nucleotides which are used as precursors for the chemical synthesis of DNA are much less polar. All the reactive sites in these synthetic molecules are chemically blocked and they are neutral, lipophilic compounds; thus, the preferred analytical and preparative mode for their separation is normal phase silica adsorption chromatography (Seliger et al., 1982).

11.1.7. Oligo- and polynucleotides

The purification of both natural and synthetic genetic material is of fundamental importance in recombinant DNA research. For example, the generation of DNA restriction fragments by initial enzymatic digestion and subsequent purification allows sequence analysis or genetic recombination to be carried out. In addition, the preparative separation of specific messenger RNA fractions is of particular value for the production and cloning of complementary DNA sequences. Also, with recent advances in the ability to chemically synthesise nucleic acids, the purification of oligonucleotides from chemical side products is essential to ensure accuracy of their function.

The predominant feature of nucleic acids in aqueous solution is the ionic nature of their phosphate groups which makes these molecules ideal candidates for separation by ion-exchange chromatography. Reversed phase separations may also be carried out by utilising the weak hydrophobic nature of the nucleic acid bases. Natural nucleic acids occur in many different size classes and group separations (e.g. between ribosomal RNA and transfer RNA) by size exclusion can also be effective. The relative merits of these methods are described in the following sections.

11.1.7.1. Anion-exchange HPLC

The separation of nucleic acids by using strong anion-exchange stationary phases has become very popular (Gait et al., 1982; McLaughlin and Romanuik, 1982). The separations depend on the charge of

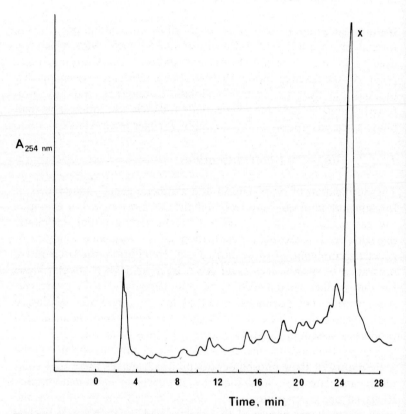

Fig. 11.1.11. Separation of oligonucleotides from a chemical synthesis mixture. Chromatographic conditions: column, Partisil 10 SAX (500×9.4 mm); mobile phase, a linear gradient from 1 M to 4 M triethylammonium acetate, pH 4.9 over 30 min; flow rate, 3.0 ml/min; temperature, 50 °C; detection, UV at 254 nm. Peak X corresponds to an oligonucleotide which is 19 nucleotides long. The peaks with a shorter retention time correspond to oligonucleotide side products.

the molecules and their interaction with the tertiary ammonium groups attached to the silica support. This mode is particularly useful for the resolution of small molecules (less than 21 nucleotides long), since the relative resolution of n and $n + 1$ molecules decreases with increasing lengths (n). Hydrophobic forces between the solute and the tertiary ammonium spacer arm of the stationary phase can influence separation but may be minimised by the addition of an organic solvent to the aqueous buffer eluent. However, it should be noted that the recovery of longer molecules from strong anion-exchange columns is relatively low because of irreversible binding to the stationary phase.

Mobile phases containing phosphate buffers have often been used in the preparative purification of oligodeoxynucleotides. However, the necessity to subsequently desalt the product limits the use of such mobile phases; furthermore, problems associated with intramolecular and intermolecular interactions (hydrogen bonding) can occur, although these can largely be overcome by including formamide in the eluent (Newton et al., 1983). An alternative mobile phase is aqueous triethylammonium acetate which can be removed by lyophilisation and in which these molecular interactions are limited (Fig. 11.1.11).

The in situ surface modification of silica with polyethyleneimine has been reported and provides resolution of oligonucleotides up to chain length 35 (Pearson and Regnier, 1983). The introduction of microparticulate ion-exchangers with a large pore size has extended the application of ion-exchange HPLC to much larger molecules, such as DNA restriction fragments of up to 500 nucleotides in length (Kato et al., 1983a).

A comparison of anion-exchange, reversed phase and reversed phase ion-pair modes for the separation of small oligonucleotides concluded that anion-exchange provides the best resolution although column life-time may be shorter (Haupt and Pingoud, 1983).

11.1.7.2. Reversed phase HPLC

The retention of oligonucleotides on reversed phase stationary phases is dominated by the hydrophobic interactions between the nucleotide bases and the stationary phase. Since these interactions are relatively independent of the specific nucleotide base, retention is determined by the relative charge, with the more highly charged molecules eluting first. The reversed phase mode has been used for the separation of ribo- and deoxyribo-oligonucleotides or their 5′-phosphorylated derivatives (Fritz et al., 1978; McFarland and Borer, 1979; Delort et al., 1984). Typically, a linear acetonitrile or methanol gradient from 5% to 30% in a weak aqueous buffer such as triethylammonium acetate (pH 7.5) may be utilised (Schott et al., 1983).

The reversed phase mode may be exploited to purify chemically synthesised oligonucleotides by deliberately leaving a labile lipophilic protecting group on the required oligonucleotide which is thus strongly retained and can be separated from the more polar reaction side products. Subsequently, the natural structure can be regenerated by chemically removing the lipophilic protecting group (Ike et al., 1983).

11.1.7.3. Reversed phase ion-pair HPLC

In reversed phase ion-pair chromatography the negative charges on the phosphate groups of the oligonucleotides are neutralised by the addition of positively charged alkylammonium ions and chromatography can be carried out on ODS stationary phases (Jost et al., 1982). This method has not been widely employed but has been used for the chain length determination of oligonucleotides after initial enzymatic digestion with a phosphodiesterase (Crowther et al., 1982).

11.1.7.4. Size exclusion HPLC

In general, high performance size exclusion chromatography is not used for the separation of oligonucleotides containing less than 100 bases owing to its low resolving power compared with either ion-exchange or reversed phase modes (Molko et al., 1981). However, size exclusion is used for the separation of much larger molecules such as transfer RNAs, ribosomal RNAs and DNA restriction fragments, although polyacrylamide gel electrophoresis is often preferred for these separations (Graeve et al., 1982).

Proteins, peptides and amino acids

11.2.1. Introduction

HPLC has made an enormous impact in the area of protein chemistry during the last decade and an annual conference specifically devoted to this topic has been inaugurated. Throughout this period commercially available stationary phases for both reversed phase and ion-exchange chromatography of large proteins (> 20 kDa) have been introduced, but the full extent of their application remains to be realised (Regnier and Gooding, 1980). Several reviews on the applications of HPLC to proteins have been published (Horvath, 1983; Henschen et al., 1985).

11.2.2. Proteins

11.2.2.1. Introduction

The recent interest in the chromatography of proteins has been markedly stimulated by the developments in recombinant DNA technology. These developments have facilitated the detailed studies of the structure of proteins (some of therapeutic potential) and introduced the possibility of redesigning the structure to alter the biological activity. Highly purified protein preparations are required for a detailed analysis using either X-ray crystallography or two-dimensional NMR.

The purification of proteins has traditionally been regarded as a difficult, laborious, time-consuming task. This is because no two proteins can be purified to the same degree by the same method and consequently each purification process has to be derived empirically. The following section is intended as a guide for researchers who are considering or who are already using HPLC for the purification of proteins and provides guidelines for the design of a successful strategy for purification.

11.2.2.2. Methods for purification

11.2.2.2.1. Physicochemical aspects. Purification strategies are usually based on a combination of separation methods which exploit the physicochemical aspects of proteins, i.e. their size, charge and hydrophobicity. If possible, an initial extraction using either organic solvent or acid can be useful, but care must be taken to avoid denaturation and modification (e.g. deamination of glutamine and asparagine). It is common practice in protein purification to fractionate the sample initially using size exclusion chromatography; however, this has the disadvantage of increasing the sample volume and therefore on a preparative scale this procedure is not recommended as a first stage. A more useful strategy is to use an initial adsorption step to concentrate the sample, which, when dealing with nanomole quantities of proteins, is a prerequisite for the development of any subsequent steps. It is also important to develop an intermediate step which allows the protein to be maintained in a stable form for long enough to accumulate sufficient quantities of protein for the development of subsequent steps.

The technique of ion-exchange chromatography is widely used for the purification of proteins since the chromatographic conditions are compatible with the stability of most proteins. Furthermore, the sample capacity is high and the selectivity is better than in any other chromatographic mode. Separations may be carried out at either high pH values (anion-exchange), or low pH values (cation-exchange). The sample can be eluted by alteration of either the pH of the mobile phase or, more commonly, by increasing the ionic strength. The behaviour of a protein on an ion-exchange stationary phase can sometimes be predicted using the technique of two-dimensional isoelectric focussing. This technique works well for hydrophilic proteins but the chromatographic behaviour of more hydrophobic proteins can be quite different from that seen under the influence of an electric field in isoelectric focussing. The inclusion of a low concentration of an organic solvent in the mobile phase (10–20%) is a common procedure in ion-exchange chromatography, and the use of high concentrations of urea (3–6 M) is also popular when dealing with samples that are sparingly soluble in aqueous media, though the latter may result in modification of lysine and possibly cysteine residues. The availability of new ion-exchange stationary phases, e.g. polymer-based (TSK-PW, Toyo Soda; FPLC, Pharmacia) and various silica-based stationary phases (Synchrom, Brownlee), makes this technique available for analytical and preparative HPLC.

After ion-exchange chromatography the sample is usually in a mobile phase containing a high salt concentration. Traditionally the salt has been removed by precipitation using ammonium sulphate followed by dialysis. However, the sample can also be applied directly to a hydrophobic stationary phase and eluted using either water or polyols; reversed phase stationary phases (wide-pore alkyl silica-based) or hydrophobic stationary phases (TSK-PW, phenyl) are ideal for this purpose.

Size exclusion chromatography can then be used either to change the mobile phase into one that is volatile and is therefore suitable for lyophilisation, or to remove other unwanted low molecular weight contaminants.

A combination of these separation methods is still very often the only approach that is available; however, recent advances have been made which greatly facilitate the task of purification of proteins.

11.2.2.2.2. Novel chromatography.   The introduction of affinity chromatography (Chapter 9) was a major advance in the purification of proteins. In contrast to the other methods, which simply exploit the physicochemical differences between different proteins, this technique also exploits certain structural and functional differences. The only disadvantage of this technique is the relatively low sample capacity of the stationary phase when a ligand of a high molecular weight is used, although this is only a problem at the preparative scale.

11.2.2.2.3. Protein engineering.   Recent advances in genetic engineering have created the possibility of altering the structure of proteins to facilitate their purification. By manipulating the genes which code for a specific protein, fusion peptides can be attached at the N and C terminus which assist their purification. The peptide extensions are subsequently removed and the protein recovered by conventional methods of purification (Shine et al., 1980). One particular method utilises the C-terminus of the protein to drastically alter the isoelectric point by the addition of a number of arginine residues (Sassenfeld and Brewer, 1984). In contrast to many of the other 'purification fusions' which attach ligands to the N-terminus, this allows the protein to fold correctly. After an initial purification by cation-exchange chromatography, the arginine residues can be removed by an exopeptidase (carboxypeptidase B), thus altering the pI of the protein of interest, and a second cation-exchange step is then carried out. The protein will elute in a different position in this second stage, whereas the contaminants will remain unaltered (unless they too possess a string of arginine residues at the C-terminus). This methodology offers the exciting prospect of designing proteins to suit pre-planned purification procedures. Perhaps in the future it may be possible to use a single process for all proteins (provided the primary sequence is known in advance).

The following sections discuss the application of HPLC to the various methods used in the isolation and analysis of proteins, peptides and amino acids. The development of new stationary phases coupled with instrumentation which allows unattended gradient development has transformed the task of purification. Some of these aspects will also be discussed in this section.

11.2.2.3. Ion-exchange chromatography

Ion-exchange chromatography (IEC) is the most widely used mode for the separation of proteins, since the optimum chromatographic conditions are compatible with the maintenance of the secondary and tertiary structure of complex biopolymers. The retention of a solute in IEC can be controlled by altering the ionic strength, pH and temperature. In addition it has been observed that these parameters can also influence the recovery (Frolik et al., 1982). However, mobile phase selection is empirical for most proteins (Kopaciewicz and Regnier, 1982) and the ability to develop chromatographic conditions using run times of less than one hour (including re-equilibration) is therefore highly beneficial. High performance anion- and cation-exchangers based on either silica or an organic polymer are now widely used and they possess a degree of stability not found in earlier types of stationary phase (Roumeliotis and Unger, 1979).

The stationary phases used for IEC of proteins are comprised of spherical beads which possess a large pore size (30–50 nm). The large

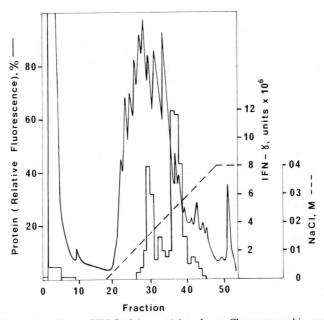

Fig. 11.2.1. Ion-exchange HPLC of (gamma) interferon. Chromatographic conditions: stationary phase, Mono S cation-exchange column (50×5 mm I.D.); mobile phase, 10 mM sodium phosphate, pH 7.0, 20% ethylene glycol, elution was achieved with a linear gradient of sodium chloride (0–400 mM) as shown; flow rate, 0.5 ml/min; temperature, ambient. Fractions of 1 ml were collected and assayed for activity. Reproduced from Freidlander et al. (1984), with permission.

pore size reduces mechanical strength and therefore lower operating pressures are required (typically no greater than 500 psi at 1 ml/min). The most popular and successful types of stationary phase for IEC of proteins are the polymer-based TSK-PW from Toyo Soda, the FPLC Mono Beads from Pharmacia, and the silica-based Synchrom beads from Synchrom Inc. The polymer-based stationary phases can be used over a wider range of pH (pH 2.0–pH 10.0) since they do not suffer from the dissolution problems associated with silica at high pH values (> pH 8.0). These stationary phases are more popular for this reason. Preparative separations on the new TSK-5PW-SP stationary phase (500 ml bed volume) have given results comparable to those obtained on an analytical size column (Brewer et al., 1986), thereby allowing scale-up for the preparation of gram quantities of proteins.

An intermediate stage in the purification of human gamma interferon from tissue culture using cation-exchange is shown in Fig. 11.2.1. A Mono-S stationary phase equilibrated in 10 mM sodium phosphate, pH 7.0 was used with 20% ethylene glycol in the equilibrating buffer. Elution was achieved with a linear salt concentration as shown.

11.2.2.4. Reversed phase chromatography

The introduction of wide-pore high performance reversed phase supports for the chromatography of large proteins (> 30 kDa) has encouraged the use of these stationary phases for proteins of all sizes (Lewis et al., 1980; Pearson et al., 1982). In theory, the wide-pore stationary phases are only necessary when the molecules are sufficiently large to be excluded from entering smaller pores. Generally, it is beneficial to use the smaller pore size stationary phases since they have a greater efficiency due to the increase in surface area available for binding. Thus, for a small protein, a higher plate number (N) will be achieved on a smaller pore size stationary phase. Although many biologically active proteins may be denatured by organic solvents or by the low pH under which reversed phase normally operates (Luiken et al., 1984), other proteins have been found to be quite stable due to their ability to re-fold upon removal of the mobile phase (Rubinstein et al., 1980). In certain circumstances the maintenance of biological

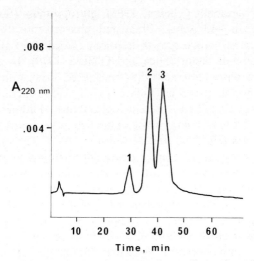

Fig. 11.2.2. Separation of native types I and III collagen. Chromatographic conditions: stationary phase, C18 Bakerbond wide-pore column (250 × 4.6 mm I.D.); mobile phase, 0.05 M ammonium bicarbonate adjusted to pH 3.2 with trifluoroacetic acid (Buffer A), tetrahydrofuran (Buffer B), stepwise gradient of 20% Buffer B for 10 min, a linear gradient of 20–30% Buffer B over 30 min and 30% B for 15 min; flow rate, 1 ml/min; temperature, 21°C. Peaks: 1, artifact; 2, type III collagen; 3, type 1 collagen. Reproduced from Smolenski et al. (1983), with permission.

activity may not be necessary, particularly if the protein is being purified for sequence analysis (Mahoney and Hermodson, 1980). An example of a reversed phase protein separation is shown in Fig. 11.2.2, where the complete separation of the mature forms of Types I and III collagen (300 kDa) from bovine skin has been achieved in less than 30 min on a wide-pore C_{18} stationary phase using a step-wise gradient of acetonitrile in trifluoroacetic acid (Smolenski et al., 1983b). Denaturation at 56°C for 30 min prior to chromatography allowed the individual subunit chains to be resolved.

There is a long list of proteins which have been successfully isolated using reversed phase chromatography despite being present in sub-nanomole quantities in vivo. These include structural proteins such as collagen, blood proteins (Kato et al., 1980; 1982), enzymes

(Ui, 1981), hormones (Voelter, 1985), glycoproteins (Stickler et al., 1982; Newman and Kahn, 1983) and pharmaceutically interesting proteins such as human growth hormone (Goeddel et al., 1979) and epidermal growth factor (Nice and O'Hare, 1979). In addition, reversed phase chromatography has been used to examine conformational changes in proteins (Luiken et al., 1984), identify peptides which contain cysteine in protein hydrolysates (Fullmer, 1984), and examine differences in chromatographic behaviour between small and large molecules (Ekstrom and Jackobson, 1984). The mechanism of retention of proteins in reversed phase chromatography is not clearly defined, but the ability to use shorter columns with no loss in resolution (O'Hare et al., 1982), suggests that the process is not controlled by simple partition chromatography alone. Therefore when assessing column efficiency it is important to use a standard which consists of a mixture of proteins with a range of hydrophobicities rather than simple organic molecules (Smolenski et al., 1984). In addition, this means that short-column systems (Brownlee cartridge type) may be most suitable for protein separations since they will allow a substantial saving in cost.

11.2.2.5. Normal phase chromatography

Normal phase chromatography (NPC) has received little attention for the purification of proteins, though there is an increasing awareness of its potential application. The reason for the disregard of this technique is that many proteins are generally considered to be insoluble in mobile phases containing a high percentage of organic solvent; conditions which are normally used in NPC. However, it is worth noting that many proteins are soluble in a relatively high concentration of organic solvent (after reversed phase chromatography) and therefore may be directly applied to NPC. Generally, NPC can be used for hydrophobic proteins which are partially soluble in organic solvents such as hexane or dimethyl sulphoxide (e.g. membrane proteins). Typically, the organic solvent content of the aqueous mobile phase is decreased in order to increase the polarity of the mobile phase. The

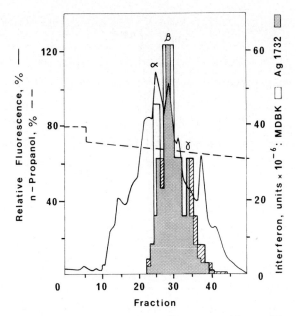

Fig. 11.2.3. HPLC of leucocyte interferon. Chromatographic conditions: column, Lichrosorb Diol (250×4.6 mm I.D.); mobile phase, 80% *n*-propanol/0.1 N sodium acetate, pH 7.5, elution was achieved with a linear gradient of decreasing propanol concentration as shown; flow rate, 0.25 ml/min; temperature, ambient; detection, post-column fluorescence using fluram. Reproduced from Rubinstein et al. (1980), with permission.

separation of interferons on a Lichrosorb diol phase demonstrates this principle (Fig. 11.2.3).

11.2.2.6. Affinity chromatography

High performance liquid affinity chromatography (HPLAC) is a recently introduced form of HPLC and few commercial stationary phases are currently available: the Beckman Fast Affinity stationary phase, stationary phases from LKB and Biorad and the Pierce High Performance Affinity stationary phase. Only a few reports are available of their use and in the past most affinity stationary phases have been prepared by proprietary methods in individual laboratories. An

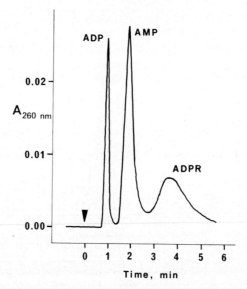

Fig. 11.2.4. HPLAC of adenine nucleotides. Chromatographic conditions: column, horse-liver alcohol dehydrogenase immobilised on silica (1000A, 10 μm); mobile phase, sodium phosphate buffer (0.25 M), pH 7.6, elution was achieved using AMP (30 μM), ADP (15 μM) and ADPR (60 μM); flow rate, 1 ml/min; temperature, ambient. Reproduced from Nilsson and Larsson (1983), with permission.

elegant demonstration of the capability of HPLAC in the biospecific purification of lactate dehydrogenase in the presence of 0.1 M pyruvate has been demonstrated using immobilised adenosine monophosphate. Initial elution of alcohol dehydrogenase can be achieved using NAD and 0.1 μM pyrazol (Lowe et al., 1981). Conversely, proteins can be coupled to high performance stationary phases (without loss of activity) and used in the preparation and resolution of nucleotide cofactors (Nilsson and Larsson, 1983) (Fig. 11.2.4). The dissociation constants for each cofactor compared favourably with those achieved from solution studies, indicating that the coenzyme binding sites were not impaired. An investigation into the purification of leukocyte interferon indicated that comparable characteristics are obtained for stationary phases based on either silica or agarose to which immunoglobulins are attached (Roy et al., 1984). Ligand 'bleeding' has been a major problem in affinity chromatography on soft gels due to physical and chemical forces but the use of silica or polymer microspheres and improved coupling methods (Chapter 9) should overcome this problem.

11.2.2.7. Size exclusion chromatography

The recent introduction of size exclusion chromatography (SEC) supports which have been specifically designed for proteins have encouraged many biochemists to make use of them (Ohno et al., 1981; Montelaro et al., 1981); however, a common finding is that many proteins behave abnormally due to either ionic or hydrophobic interactions with the stationary phase. At present, ideal conditions for the separation of each protein can only be determined empirically, although general guidelines can be found in Chapter 5. Apart from the use of SEC in purification and desalting, the most popular application

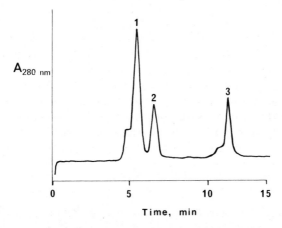

Fig. 11.2.5. HPLC separation of reduced and alkylated human IgG into its subunits. Chromatographic conditions: column, Bio-Sil TSK-250 (300×7.5 mm I.D.); mobile phase, 0.02 M sodium phosphate, pH 6.5, 5.0 M guanidine HCl; flow rate, 1.0 ml/min; temperature, ambient. Peaks: 1, heavy chain (51 kDa); 2, light chain (26 kDa); 3, alkylating reagent (iodoacetamide). Reproduced from Biorad Information Catalogue, with permission.

is for size determination under strong denaturing conditions (Fig. 11.2.5). SEC can also be used for the detection of high molecular weight aggregates when assessing the purity of a sample. The mobile phases used in SEC need to contain additives to prevent chemical interaction with the stationary phase but which do not disrupt the aggregates, at the same time, e.g. non-ionic detergents, propylene glycol or mild chaotropic salts. (A comprehensive list of applications of high performance SEC is available from Toya Soda Manufacturing Co. Ltd.)

11.2.3. Peptides

11.2.3.1. Introduction

HPLC has been used for the analysis and purification of peptides for many years, both in the production of synthetic peptides and in the isolation of naturally occurring peptides. An extensive literature exists on the use of the technique, and the examples quoted here serve only to illustrate either recent developments, or specific and novel applications. In addition, the application of the new ion-exchange stationary phase in the separation of peptides is well documented in literature available from the manufacturers. Application of HPLC to purification, characterisation (peptide mapping), sequencing and purity assessment is now universally recognised. In addition, successful assays based on HPLC have been developed.

11.2.3.2. Ion-exchange chromatography

The extensive use of reversed phase ion-pair HPLC in peptides has meant that some aspects of ion-exchange HPLC are not being exploited, e.g. for peptide mapping. The isocratic separation of the two desired components from a tryptic digest of an epidermal factor fusion peptide can be achieved using a cation-exchange stationary phase in combination with a step-wise gradient of sodium chloride in

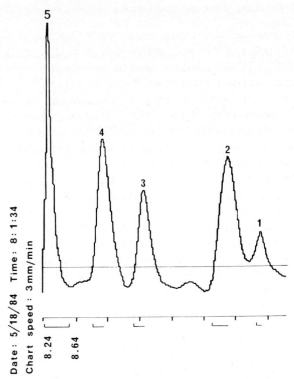

Fig. 11.2.6. Separation of a limited tryptic digest of poly-arg urogastrone. Chromato-graphic conditions: column, TSK-SP 5PW (150×7.5 mm I.D.); mobile phase, 50 mM sodium acetate, 100 mM sodium chloride pH 3.6 (Buffer A), 500 mM sodium acetate, 1 M sodium chloride, pH 3.6 (Buffer B) 35%; Buffer B for 10 min, 50% Buffer B for 10 min, 25% Buffer B for 15 min; flow rate, 1.0 ml/min. Peaks: 1, injection; 2, native urogastrone; 3,4, partial digestion products; 5, poly-arg-urogastrone.

50 mM acetate buffer, pH 4.0 (Fig. 11.2.6). The purity of each species is greater than 90%. In contrast to reversed phase separations, the sample is eluted in a non-volatile buffer which makes subsequent analysis more difficult. However, ion-exchange chromatography does not produce some of the artifacts sometimes found in reversed phase chromatography due to denaturation.

11.2.3.3. Reversed phase chromatography

The extensive use of reversed phase HPLC for the separation of peptides (Voelter, 1985) has resulted in a method being devised for the prediction of the retention time of an unknown peptide whose amino acid composition is known. Based on a study of the retention of a series of peptides, retention coefficients can be ascribed to individual amino acids which in turn can be related to their hydrophobicity. The sum of the retention coefficients of the amino acids in the unknown peptide can be used to predict their elution position (Meek and Rosetti, 1985). The absolute values of the retention coefficients depend on the composition of the mobile phase. However, peptides larger than 10 kDa may behave anomalously, probably due to conformational changes caused by organic solvent and pH. Trifluoroacetic acid is frequently used in the mobile phase (Mahoney and Hermodson, 1980) because of its good UV characteristics and its ability to solubilise many peptides. Other mobile phases are derived from well-known buffers though many cannot be used in certain circumstances. For example, acetate buffers cannot be used at high sensitivity below 230 nm.

The use of RPC as an adjunct to protein sequencing has gained universal acceptance for the isolation of peptide fragments after enzymatic digestion of a protein (Tempst et al., 1984; Hunkapiller and Hood, 1978). Each peptide can be processed by reaction with the Edman reagent and the sequence obtained by reversed phase chromatography of PTH–amino acids (Section 11.2.4). Another common use of RPC is in the assessment of the purity of peptides, although it is becoming clear that for larger peptides more than one buffer system should be used to ensure resolution of all the minor impurities. UV detection is wholly unsatisfactory for some peptides (since detection depends upon the presence of tyrosine, tryptophan and phenylalanine) and therefore extensive development has been carried out using post-column fluorescence detection (Section 11.2.4).

11.2.3.4. Size exclusion chromatography

At present the resolution of peptides and proteins smaller than 10 kDa using SEC is not adequate and this mode of chromatography is

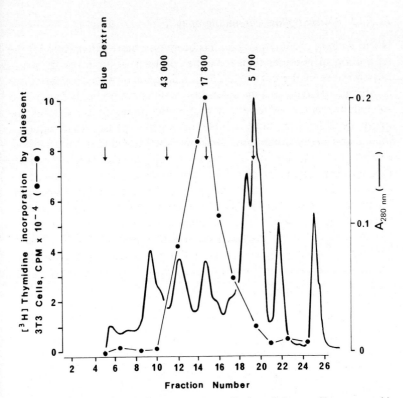

Fig. 11.2.7. Size exclusion HPLC of human milk growth factor. Chromatographic conditions: column, TSK 2000SW (60×7.5 mm I.D.); mobile phase, 0.05 M potassium phosphate, 0.1 M ammonium sulphate, pH 7.0; flow rate, 0.8 ml/min. The column was previously calibrated with markers of known molecular weight, indicated by arrows. Reproduced from Sullivan et al. (1983), with permission.

therefore used mainly for bulk separations of peptides from large molecular weight material, e.g. during the preparation of tryptic fragments, or for crude fractionations (Sullivan et al., 1983) (Fig. 11.2.7). Another application is in the recovery of peptides after chemical modification, e.g. to remove the alkylating agent after alkylation (Fig. 11.2.5). Accurate molecular weight determination, even with a narrow range of standards, is not reliable unless carried out in the presence of strong denaturants.

11.2.3.5. Normal phase chromatography

Normal phase chromatography has been used more extensively for the separation of peptides than for either proteins or amino acids, primarily because of the poor solubility of proteins in high concentrations of organic solvents and the solubility of most amino acids in aqueous solvents (except a few of the hydrophobic amino acids). Conversely, many peptides are more soluble in organic phases than aqueous phases and many subtle interactions can be identified by this method. In Fig. 11.2.8 the elution of a series of peptides differing only in the position of a single glycine residue can be seen. The interaction of the glycine residue varies with its position in the chain and influences the retention (Naider et al., 1983). A common problem which is encountered during the synthesis or sequencing of proteins and peptides is the production of an intermediate which is not soluble in aqueous conditions; this may be overcome by solubilising in aqueous dimethylsulphoxide and then exploiting normal phase chromatography.

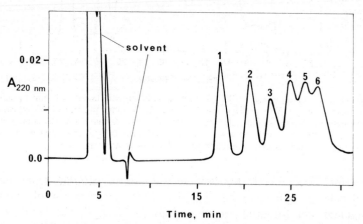

Fig. 11.2.8. Normal phase HPLC separation of six isomeric hexapeptides. Chromatographic conditions: stationary phase, μPorasil silica column (300×3.9 mm I.D.); mobile phase, cyclohexane–isopropanol–methanol ($92:6:2$); flow rate, 1.0 ml/min; temperature, ambient. Peaks: 1, Boc-Met_3-Gly-Met_2-OMe; 2, Boc-Met_2-Gly-Met_3-OMe; 3, Boc-Met_5-Gly-OMe; 4, Boc-Gly-Met_5-OMe; 5, Boc-Met_4-Gly-Met-OMe. Reproduced from Naider et al. (1983), with permission.

11.2.4. Amino acids

11.2.4.1. Introduction

The development of high performance stationary phases for amino acid analysis has been vigorously pursued since the introduction of the first amino acid analyser (Spackman et al., 1950). The low UV extinction coefficient of most amino acids means that current detection methods depend upon the reaction of the primary amino group of the amino acid to yield a coloured or fluorescent derivative. The most popular reagents, referred to by their most common nomenclature, are the following:

Ninhydrin
Orthophthalaldehyde (OPA)
Fluorescamine (Fluram)
Phenyl isothiocyanate
Dansyl chloride
Dabsyl chloride
Diaminoazobenzeneisothiocyanate
UV at 200 nm

Other reagents are available but have not yet found widespread acceptance. For example, fluorenylmethyl chloroformate, which is commonly used as a blocking agent in peptide synthesis, can be used in pre-column labelling with sensitivity in the femtomole range (Varian).

11.2.4.2. Ion-exchange chromatography

The majority of the commercially available amino acid analysers employ an ion-exchange resin for the separation of amino acids with ninhydrin as the detection reagent (Fig. 11.2.9), although OPA and fluram can also be utilised. Despite recent developments (Section 11.2.4.3), such a dedicated instrument is often preferred in routine analysis for reasons of ease of sample handling and reproducibility.

11.2.4.3. Reversed phase chromatography

The increasing popularity of RPC has prompted many researchers and commercial laboratories to develop separations which can be

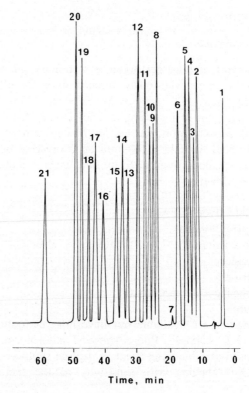

Fig. 11.2.9. Amino acid analysis using a conventional analyser. Chromatographic conditions: column, Dionex DC-6A (300 × 4.6 mm I.D.); mobile phase, sodium citrate, three-buffer programme; detection, post-column reagent ninhydrin (absorbance 1 full scale). Peaks: 1, cysteic acid; 2, aspartic acid; 3, methionine sulphone; 4, threonine; 5, serine; 6, glutamic acid; 7, proline; 8, glycine; 9, alanine; 10, half cystine; 11, valine; 12, methionine; 13, isoleucine; 14, leucine; 15, *N*-leucine; 16, tyrosine; 17, phenylalanine; 18, ammonia; 19, lysine; 20, histidine; 21, arginine. Reproduced from Beckman information catalogue, with permission.

carried out using conventional HPLC equipment. However, the separations reported using RPC do not encompass all of the capabilities of the conventional analyser (particularly when analysing physiological fluids), but considering the intense interest in RPC it may only be a matter of time before equivalent separations are achieved. Many detection procedures have been specifically developed for use in conjunction with reversed phase systems and these are discussed below.

11.2.4.3.1. Pre-column derivatisation

Orthopthalaldehyde (OPA). The simplicity of the chemistry of the reaction of OPA with amino acids has led to the widespread use of this reagent and some excellent separations have been reported. The separation times of OPA–amino acids have been dramatically reduced by decreasing the column length and increasing the flow rate and even faster analysis can be achieved using a guard column alone packed with the same support (Jones and Gilligan, 1983). It is now possible to automate this procedure using an auto sampling device to derivatise the samples prior to chromatography. Disadvantages of OPA derivatives include their relatively short half-life compared with fluram and dansyl derivatives and the poor reactivity against secondary amines (the addition of hypochlorite can offset this latter problem).

11.2.4.3.2. Post-column derivatisation

Orthopthalaldehyde (OPA). To achieve post-column derivatisation with OPA a reaction coil (4 mm × 0.3 mm I.D.) is usually inserted in-line after the point of mixing with the fluorogenic reagent. Since the dead volume of the coil contributes to band broadening effects, the solution is not as good as with the pre-column system. Slower flow rates are required to resolve certain residues which thereby increases the analysis time. However, the main advantages of OPA derivatisation (which is also used in conventional analysers) are the reproducibility and the fact that, unlike ninhydrin, a heating device is not required for the reaction.

Fluorescamine (Fluram).   Fluorescamine was developed from the study of ninhydrin chemistry as a reagent for detecting amino groups (Udenfriend et al., 1972) and has been successfully used in the reversed phase chromatographic analysis of proteins and peptides using an on-line post-column detection system. This reagent requires two separate pumps for the delivery of the mobile phase buffer and the reagent and this method may therefore be considered to be less

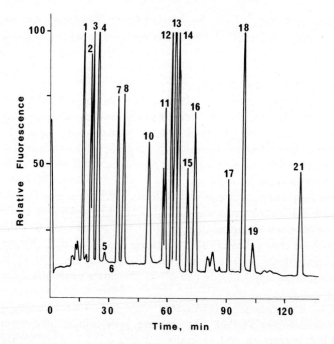

Fig. 11.2.10. HPLC of human leucocyte interferon hydrolysate using post-column fluorescence detection with fluorescamine. Chromatographic conditions: column, DC-4A (Durrum) (500×4.6 mm I.D.); mobile phase, 0.175 N sodium citrate, 2.5% isopropanol, pH 3.5 (Buffer A), 0.175 N sodium citrate, pH 3.45 (Buffer B), 0.2 N sodium citrate, 0.05% thiodiglycol, pH 4.1 (Buffer C), 0.52 N sodium citrate, 1 N sodium chloride, 0.05% thiodiglycol, pH 7.9 (Buffer D), elution was achieved with the following gradient: Buffer A (15 min), Buffer B (17 min), Buffer C (24 min), Buffer D (41 min), flow rate, 18 ml/h; temperature, 57°C. Peaks: 1, aspartic acid; 2, threonine; 3, serine; 4, glutamic acid; 5, cysteine; 6, proline; 7, glycine; 8, alanine; 10, valine; 11, methionine; 12, isoleucine; 13, leucine; 14, norleucine; 15, tyrosine; 16, phenylalanine; 17, histidine; 18, lysine; 19, ammonia; 21, arginine. Reproduced from Stein and Brink (1981), with permission.

facile than using OPA. The level of sensitivity is similar for both reagents and is over 10 times better than for ninhydrin. Fluram also reacts poorly with secondary amines (e.g. proline), but this can be overcome by the addition of succinic anhydride (Stein and Brink, 1981). Fig. 11.2.10 shows a separation using ion-exchange chromatography.

Phenylisothiocyanate (PITC).   The use of PITC (Edman Reagent) to form thiohydantoin (PTH) derivatives of amino acids for protein sequencing is well known. The development of reversed phase systems for the separation of these PTH derivatives has resulted in the introduction of both isocratic and gradient elution systems (Fig. 11.2.11). A microbore HPLC unit for PTH–amino acid analysis is now available along with the recently introduced gas phase sequenator

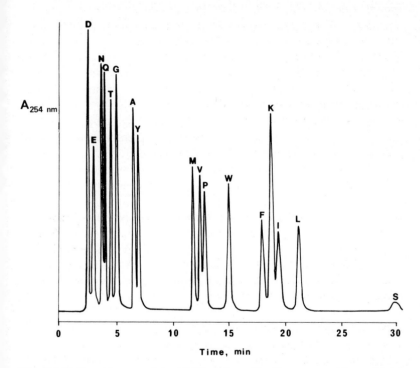

Fig. 11.2.11. Isocratic separation of PTH–amino acids. Chromatographic conditions: column, Ultrasphere ODS (250×4.6 mm I.D.); mobile phase, 0.01 M sodium acetate (pH 4.9)–acetonitrile (62.2:37.8); flow rate, 1 ml/min; temperature, ambient. Peak identity corresponding to the single letter code for amino acids: D, aspartic acid; E, glutamic acid; N, asparagine; Q, glutamine; T, threonine; G, glycine; A, alanine; Y, tyrosine; M, methionine; V, valine; P, proline; W, tryptophan; F, phenylalanine; K, lysine; I, isoleucine; L, leucine; S, serine. Reproduced from Noyes (1983), with permission.

(Applied Biosystems). Recently, a system for the separation of the less stable phenylthiocarbamyl (PTC) derivatives was published (Heinrikson and Meredith, 1984) and a commercial kit is now available (Waters, Pico-Tag) which greatly enhances not only the speed of analysis, but also increases the sensitivity of detection to the low picomole range. The system is reproducible but requires a skilled operator.

Dansyl chloride (DNS-Cl) (1-dimethylaminonaphthalene-5-sulphonyl-chloride). This reagent was originally introduced into protein chemistry for end-group analysis over twenty years ago (Gray and Hartley, 1963) and has been widely used because of the simplicity of the reaction and its ability to react with both primary and secondary amines, unlike OPA and fluram. Furthermore, in contrast to other fluorescent reagents, the dansyl derivatives are stable to acid hydrolysis, and can therefore be used in N-group labelling before hydrolysis. HPLC separations of dansyl derivatives have recently been published (Tapuhi et al., 1981). Sensitivity of detection is at the low picomole level. The sensitivity is limited because of the side-reactions which can occur with lysine and, to a lesser extent, histidine and tyrosine.

Dabsyl chloride (DBS-Cl) (4,4-dimethylaminoazobenzene-4-sulphonyl-chloride). The reaction of amino acids with dabsyl chloride produces a compound which can be detected at the picomole level using UV detection (436 nm). The product is stable at room temperature (Lin, 1984) but has not received the same attention as other reagents (Chang et al., 1984).

Diaminoazobenzeneisothiocyanate (DABITC) (4-N,N-dimethylamino-azobenzene-4-isothiocyanate). This reagent has been used successfully in the manual sequencing of proteins and peptides (Chang et al., 1980). Detection of the thiohydantoin derivatives (DABTH) can be carried out on thin layer chromatography plates with the derivatised amino acids appearing as red spots upon exposure to acid vapour. Recently, an isocratic reversed phase system with UV detection at 436 nm gave a sensitivity down to 5 picomole (Lehmann and Wittmann-Liebold, 1984).

11.2.4.4. Normal phase chromatography

The different selectivity of amino acids on normal phase supports compared with reversed phase supports allows many otherwise dif-

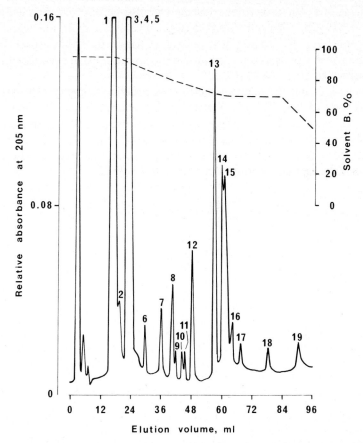

Fig. 11.2.12. Normal phase separation of amino acids. Chromatographic conditions: column, Zorbax NH_2 (250×4.6 mm I.D.); mobile phase, 10 mM potassium phosphate, pH 4.3 (A), acetonitrile–water 50:7 (v/v) (B); flow rate, 2 ml/min; temperature, 35 °C. Peaks: 1, phenylalanine; 2, leucine; 3, isoleucine; 4, methionine; 5, tyrosine; 6, valine; 7, proline; 8, alanine; 9, hypro; 10, threonine; 11, glycine; 12, serine; 13, histidine; 14, cysteine; 15, arginine; 16, lysine; 17, hydroxylysine; 18, glutamic acid; 19, aspartic acid. Reproduced from Smolenski et al. (1983), with permission.

ficult separations to be carried out (Fig. 11.2.12). For example, using a modification of a previously described method (Schuster, 1980), most of the amino acids present in a collagen hydrolysate can be separated and, more importantly, the cross-linked residues hydroxylysinonorleucine and dehydrolysinonorleucine can also be resolved (Smolenski et al., 1983b). The only disadvantage of this system is the relative instability of the amino phase column (Chapter 6). Other aspects of amino acid separations (i.e. enantiomeric separations) are discussed in Chapter 9.

The immense popularity of HPLC in protein chemistry will mean that some of the conditions for separation reported in the present text may already have been surpassed by either improved separations or new technological advances. The reader is therefore strongly recommended not only to review the current literature regularly, but also to attend the regular meetings devoted to this subject.

Lipids

There are several major classes and subclasses which comprise that group of compounds termed lipids. Lipids perform a wide variety of biological functions including: (*a*) functioning as coenzymes in enzyme reactions; (*b*) acting as structural components in membranes; (*c*) forming protective outer layers on the cell walls of some bacteria, plants and insects, (*d*) acting as metabolic fuel and (*e*) acting as a storage form for metabolic fuel. In this section we shall discuss the chromatography and detection of the lipids, including fatty acids, phospholipids, sphingolipids, ceramides, glycosylceramides, gangliosides and acylglycerols. Details of extraction procedures are not included in the text as these vary greatly depending on both the class of lipid and the source from which the lipid is extracted; for details of extraction systems the interested reader is referred elsewhere (Christie, 1982).

11.3.1. Fatty acids

Until recently, gas–liquid chromatography (GLC) was the most popular method for the identification and quantitation of fatty acids; however, with the emergence of new derivatisation techniques and improvements in detector technology HPLC offers a viable alternative. The most frequently used HPLC mode for the analysis of fatty acids is reversed phase either with or without pre-column derivatisation; additionally, argentation HPLC has been used for the resolution of certain positional and geometric isomers of underivatised polyunsaturated fatty acids.

In reversed phase HPLC the stationary phase used is almost always ODS although occasionally shorter alkyl chains have been used. The mobile phases which have been most popularly employed include methanol–water and acetonitrile–water. A good example of the resolving power of reversed phase HPLC for fatty acids is shown in Fig. 11.3.1, where six different C_{20} fatty acids were resolved on an ODS column by using two different mobile phase mixtures of methanol–water–acetic acid (Batta et al., 1984). It is apparent from the chromatograms that as the degree of unsaturation of the fatty acid was increased the retention time was decreased and also that by variation of the polarity of the mobile phase different positional isomers of the fatty acids were isolated. A gradient system, using a mobile phase containing an increasing concentration of acetonitrile in water, has also been commonly used for the resolution of fatty acids, although some problems with overlapping 'critical pairs' of fatty acids in the eluting peaks have been noted (Bussell et al., 1979). It has been demonstrated that mobile phases of acetonitrile–water in combination with an ODS stationary phase may be used to resolve both positional isomers of arachidonic acid (58% acetonitrile in water) and also geometric isomers of linoleic acid (60% acetonitrile in water) although *cis–trans* and *trans–cis* isomers were not resolved under these conditions (Van Rollins et al., 1982). The same authors noted that for a mixture of polyunsaturated fatty acids derived from brain glycerophospholipids the resolution of various fatty acids and their isomers was dependent on a specific acetonitrile concentration which was optimal for individual ODS columns supplied by different manufacturers. In general, however, as the chain length of the fatty acid was increased, the retention time was increased, and as the degree of unsaturation was increased the retention was decreased. Where problems are encountered in the resolution of adjacent species of fatty acids, a chromatographic system which may help is a LiChrosorb Hibar II RP-8 reversed phase column with a mobile phase of tetrahydrofuran–acetonitrile–water (3 : 67 : 30) containing 0.1% acetic acid (Bailie et al., 1982).

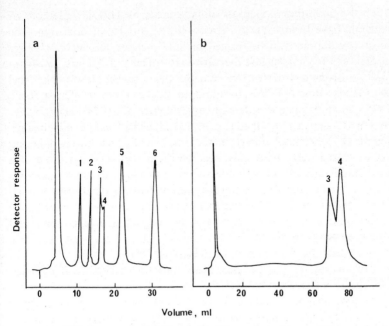

Fig. 11.3.1. HPLC separation of fatty acids. Chromatographic conditions: column, Radial-Pak C_{18} cartridge (5 μm) (100 × 8 mm I.D.) attached to a guard column (49 × 4.6 mm I.D.) dry packed with 37–50 μm C_{18} Corasil reversed phase material; mobile phases, methanol–water–acetic acid (89:11:0.2) (A), methanol–water–acetic acid (80:20:0.2) (B); flow rates, 1.0 ml/min (A), 2.0 ml/min (B); detection, UV at 214 nm. Peaks: 1, eicosapentaenoic acid (20:5, $\Delta^{5,8,11,14,17}$); 2, arachidonic acid (20:4, $\Delta^{8,11,14,17}$); 3, eicosatrienoic acid (20:3, $\Delta^{8,11,14}$); 4, eicosatrienoic acid (20:3, $\Delta^{11,14,17}$); 5, eicosadienoic (20:2, $\Delta^{11,17}$); 6, eicosaenoic acid (20:1, Δ^{11}). The peaks represent 1.25–3.75 μg of fatty acid injected on column. Reproduced from Batta et al. (1984), with permission.

The detectors most frequently used to monitor the elution of underivatised fatty acids have been differential refractive index detectors (Bailie et al., 1982; Svensson et al., 1982), moving wire flame ionisation detectors (Ozcimder and Hammers, 1980) and UV detection at low wavelengths (Van Rollins et al., 1982; Batta et al., 1984). Clearly, detection using these methods does not provide very great sensitivity at present and in instances where this is required pre-column derivatisation of the fatty acids should be utilised.

In the quantitative analysis of fatty acids by HPLC a plethora of reagents have been used to increase the sensitivity of detection. The most popular derivatives formed include: benzyl derivatives (Polizer et al., 1973); 2-naphthacyl derivatives (Cooper and Anders, 1974); o- and p-nitrobenzyl derivatives (Knapp and Krueger, 1975); phenacyl derivatives (Borch, 1975); p-bromophenacyl derivatives (Durst et al., 1975); methoxyphenacyl derivatives (Miller et al., 1978), and 1-naphthylamide derivatives (Ikeda et al., 1983). The benzyl, nitrobenzyl, phenacyl, p-bromophenacyl, methoxyphenacyl, 1-naphthylamide and 2-naphthacyl derivatives may be monitored with UV detectors at a wavelength of 254 nm and sensitivities are reported which suggest a lower limit of detection of approximately 4 ng (Cooper and Anders, 1974; Miller et al., 1978). Fluorescent detection of 2-naphthacyl derivatives of fatty acids may also be utilised with excitation and emission wavelengths of 290 nm and 450 nm, respectively, and this provides enhanced sensitivity of detection (Distler, 1980).

Chromatography of the derivatised fatty acids is generally carried out on ODS stationary phases with mobile phases of methanol–water or acetonitrile–water, although the concentration of organic modifier is generally higher than that used in the chromatography of underivatised fatty acids. The elution order of the fatty acid derivatives is the same as that observed with the underivatised species, i.e. retention time is increased as the chain length of the fatty acid is increased and decreased as the degree of unsaturation is increased.

While reversed phase chromatography is overwhelmingly the most popular technique for the separation of fatty acids and their derivatives, other chromatographic modes are utilised. For example, normal phase chromatography using a silica stationary phase with mobile phase gradient systems of chloroform–hexane mixtures have been used for the separation of p-bromophenacyl esters of the short chain carboxylic acids from arthropods (Weatherston et al., 1978). Argentation chromatography will be discussed in some detail in a later section (Section 11.5.3) and has also been used in the chromatography of fatty acids. In general, the stationary phase most commonly utilised for argentation chromatography of fatty acids is a silica column loaded with dilute solutions of silver nitrate. The mobile phase used is usually hexane to which a small percentage of a more polar organic modifier, such as acetonitrile, has been added. Argentation chromatography has been used for the resolution of geometric isomers of fatty acid methyl esters (Scholfield, 1979) and also for the resolution of positional isomers of methyl esters of linoleic acid (Battaglia and Frohlich, 1980). In general, however, it is recognised that either as a pre-fractionation method before GLC analysis or as a general method of resolution, argentation HPLC is less useful than reversed phase HPLC, except in instances where it is used to separate on the basis of the degree of unsaturation of the fatty acid (Ozcimder and Hammers, 1980).

11.3.2. Phospholipids

Since the introduction of HPLC for the isolation and quantitation of phospholipids, the vast majority of separations have utilised normal phase chromatographic techniques on silica columns with a wide variety of mobile phases. In 1975, a separation of the biphenyl-carbonyl derivatives of phosphatidylethanolamine, phosphatidylserine, lysophosphatidylethanolamine and ethanolamine plasmalogens was demonstrated using a silica stationary phase with a mobile phase of dichloromethane–methanol–15 M aqueous ammonia (92 : 8 : 1 or 80 : 15 : 3) (Jungalwala et al., 1975). This report was followed by another from the same group (Jungalwala et al., 1976) demonstrating resolution of non-derivatised sphingomyelin and phosphatidylcholine on a silica stationary phase in combination with a mobile phase of acetonitrile–methanol–water (65 : 21 : 14). The observation that inclusion of either an acid or a base in the mobile phase gives an improved separation of phospholipids on silica thin layer chromatography plates has prompted some groups to include either phosphoric acid or sulphuric acid in the HPLC mobile phase. In Fig. 11.3.2 the resolution of all the major phospholipid components from tissue lipid extracts are shown in a single chromatographic run which used a silica stationary phase with a mobile phase of acetonitrile–methanol–85% phosphoric acid (130 : 5 : 1.5, v/v). In this chromatogram the characteristic trough and peak of the co-eluting phosphoric acid and phosphatidylglycerol are shown. Under these

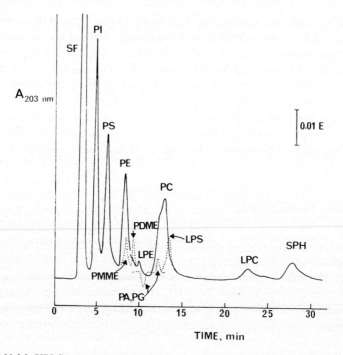

Fig. 11.3.2. HPLC separation of phospholipid standards. Chromatographic conditions: column, Micro-Pak SI-10 column (300 × 4 mm I.D.); mobile phase, acetonitrile–methanol–85% phosphoric acid (130:5:1.5, v/v/v); flow rate, 1 ml/min; temperature, ambient; detection, UV at 203 nm. Peaks: LPC, lysophosphatidylcholine (5 μg); LPE, lysophosphatidylethanolamine (5 μg); LPS, lysophosphatidylserine; PA, phosphatidic acid; PC, phosphatidylcholine (0.5 μg); PDME, phosphatidylmethylethanolamine; PE, phosphatidylethanolamine; PG, phosphatidylglycerol; PI, phosphatidylinositol (2.5 μg); PMME, phosphatidylmonomethylethanolamine; PS, phosphatidylserine (0.5 μg); SF, solvent front; SPH, sphingomyelin (2.5 μg). Reproduced from Chen and Kou (1982), with permission.

conditions the elution of cardiolipin and neutral lipids in the solvent front can interfere with the detection of phosphatidylinositol if column performance deteriorates and the authors recommend that the silica column should be washed daily with 30 ml of each of methanol–water (1 : 1), methanol and dichloromethane before storing it overnight in hexane.

The inclusion of phosphoric acid in the mobile phase increases the potential for error in subsequent quantitation by lipid phosphorus determinations and has been replaced with a mobile phase containing acetonitrile–methanol–sulphuric acid (100 : 3 : 0.05) to provide resolution of phosphatidylinositol, phosphatidylserine, phosphatidylethanolamine, phosphatidylcholine, lysophosphatidylcholine and sphingomyelin (Kaduce et al., 1983). The authors reported that a reduction in the sulphuric acid content of the mobile phase caused a broadening of the eluted peaks and an increase in the retention of phosphatidylserine, phosphatidylethanolamine and phosphatidylcholine, while if omitted, these components did not elute. It was also noted that if the methanol content of the mobile phase was increased then the retention times of all the phospholipids were decreased. Samples were therefore injected in chloroform–diethyl ether (1 : 1) to avoid altering the concentration of methanol in the mobile phase.

Several other isocratic separations of phospholipids on silica have been reported using mobile phases comprising mixtures of hexane–propanol–water (Geurts van Kessel et al., 1977; Abood et al., 1978), chloroform–propanol–acetic acid–water (50 : 55 : 2.5–5 : 5–10) (Blom et al., 1979) and chloroform–methanol (95 : 5) (Paton et al., 1983).

A number of gradient systems have been described which offer the potential for an improved resolution of the commonly occurring phospholipids. One of these gradient systems demonstrates the resolution of the recently discovered, biologically active 1-alkyl-2-acetyl-*sn*-glycerol-3-phosphorylcholine (alkylacetyl-GPC; platelet-activating factor; PAF) (Blank and Snyder, 1983). The stationary phase was silica with a mobile phase starting from 96% 2-propanol–hexane (1 : 1)–4% water, with a linear gradient to 8% water over a 15 min period. Under these chromatographic conditions, the elution order was: phosphatidylglycerol, phosphatidylethanolamine/phosphatidylinositol, phosphatidic acid, phosphatidylserine, hexadecyl-acetyl-GPC, acylacetyl-GPC, lysophosphatidylcholine. Other gradient systems used with silica stationary phases have used mobile phases of 100% propanol–ethyl acetate–benzene–water (130 : 80 : 30 : 20) to 100% propanol–toluene–acetic acid–water (93 : 110 : 15 : 15), although due to the presence of benzene and toluene this mobile phase system may not be used in conjunction with UV detection (Alam et al., 1982). An alternative system used mobile phases of 70% hexane–2-propanol (50 : 50) to 100% hexane–2-propanol–tetrahydrofuran–0.1 M NH_4Cl (40 : 45 : 5 : 10) for the resolution of a number of phospholipid classes (Yandrasitz et al., 1981). These authors suggested that the presence of NH_4Cl in the mobile phase sharpened the eluting peaks, most notably those of phosphatidylserine and phosphatidylinositol. It was also observed that a decrease in the content of 2-propanol resulted in the phospholipids eluting earlier and it was proposed that 2-propanol may act by enhancing the interaction of ionised groups with the silica surface of the stationary phase.

The use of mobile phases containing large amounts of either water or methanol for extended periods can result in problems due to the instability of the silica stationary phase. These difficulties have been overcome by employing an anion-exchange stationary phase such as μBondapak-NH_2 (Waters Assoc.) as a normal phase packing with a mobile phase gradient containing chloroform and methanol–water (25 : 1 or 25 : 4) (Kiuchi et al., 1977). The elution order of lipids from the column was: phosphatidic acid, phosphatidylglycerol, phosphatidylserine / tristearin, phosphatidylcholine. A more recent report (Hanson et al., 1981) described the use of an Ultrasil-NH_2 column (Altex) with a gradient system containing hexane–2-propanol (11 : 16) and hexane–2-propanol–methanol–water (11 : 16 : 2 : 3), which produced an elution order of: cholesterol, cerebroside, phosphatidylcholine, sphingomyelin, lysophosphatidylcholine, phosphatidylethanolamine, lysophosphatidylethanolamine. The results of Hanson et al. (1981) and Kiuchi et al. (1977) demonstrate the differences in selectivity when separating phospholipids on amine stationary phases compared with those on silica since the elution order of phosphatidylcholine and phosphatidylethanolamine is reversed on amine columns.

The observation that at neutral pH most membrane lipids contain a positively charged primary or quaternary amino group has resulted in the development of a chromatographic system using a cation-exchange column (Gross and Sobel, 1980). A mobile phase of acetonitrile–methanol–water (400 : 100 : 34) was used with a Whatman PXS 10/25 SCX cation-exchange column to separate phosphatidylethanolamine, lysophosphatidylethanolamine, phosphatidylcholine, sphingomyelin and lysophosphatidylcholine in this elution order. The limitation of this chromatographic system is that polar phospholipids which do not contain an amino functional group are eluted in the void volume due to the polarity of the solvents used.

More recently, reversed phase HPLC has also been used for the separation of both natural and synthetic phosphatidylcholines (Smith and Jungalwala, 1981; Porter et al., 1979; Compton and Purdy, 1982; Crawford et al., 1980). ODS stationary phases were used and the mobile phases included methanol–1 mM phosphate buffer, pH 7.4 (95 : 5) (Smith and Jungalwala, 1981) and chloroform–water–methanol (10 : 10 : 100) (Porter et al., 1979). In these reports the phosphatidylcholines were resolved into molecular species, the retention being dependent on the carbon number and degree of unsaturation of the fatty acid side-chains. In general, retention time decreased as the fatty acid chains became shorter and as the number of double bonds increased. A recent report describes the separation of all the major glycerophospholipids into molecular species using an Ultrasphere ODS column (Altex) with a mobile phase of 20 mM choline chloride in methanol–water–acetonitrile (181 : 14 : 5) (Patton et al., 1982).

One final word of advice for the potential chromatographer who intends to separate and quantitate phospholipids from tissue extracts using HPLC with any of the chromatographic columns described earlier is to use a guard column to enhance the lifetime of the analytical column.

The most popular detection system used in the HPLC of phospholipids uses UV absorbance at wavelengths between 200 and 206 nm. The absorbance of lipids at this wavelength is mostly due to the presence of carbon–carbon double bonds, although other groups may contribute to a lesser degree.

While UV detection is both sensitive and non-destructive, the precise quantitation of lipids is complicated since the varying degrees of unsaturation of lipids gives rise to large variations in extinction coefficients for different phospholipid classes. This problem may be overcome in two ways:

(a) The eluting peaks may be collected and subjected to lipid phosphorus analysis.

(b) An apparent extinction coefficient may be obtained for a standard sample which, together with the area under a given eluting peak, can provide an estimate of the lipid present (Jungalwala et al., 1976). The sensitivity of UV detection is of the order of 1 nmol of phosphatidylcholine providing that at least one double bond is present per molecule (Jungalwala et al., 1976).

An alternative strategy to increase the sensitivity of detection of phospholipids is to form a derivative with an enhanced UV absorbance. One highly sensitive method for the quantitative measurement of phospholipids containing primary amino groups is to form biphenylcarbonyl derivatives (Jungalwala et al., 1975). This method has been applied to the detection of phosphatidylserine, phosphatidylethanolamine and lysophosphatidylethanolamine, where the lower limit of detection was suggested to be 10–13 pmol or 0.3–0.4 ng of phospholipid phosphorus (Jungalwala et al., 1975).

Similarly, phospholipids containing primary amino groups can be derivatized with DNS-Cl (1-dimethylaminonaphthalene-5-sulphonyl chloride). DNS derivatives are highly fluorescent and, using an excitation wavelength of 342 nm and emission wavelength of 500 nm, a lower limit of detection of about 20 pmol has been achieved (Chen et al., 1981).

Two other detection systems which have been used for the quantitation of phospholipids are flame ionisation detectors (Kiuchi et al., 1977) and refractive index (RI) detectors (Paton et al., 1983). At the time of writing no flame ionisation detectors for use with HPLC systems are commercially available. The limit of detection using RI detectors is dependent upon the specific phospholipid, but has been reported to vary in extracts of amniotic fluid between 0.6 μmol/l for phosphatidylethanolamine to 55 μmol/l for lysolecithin. It is to be anticipated that technical advances will provide considerable increases in the limit of sensitivity of RI detectors.

Where radioactively labelled phospholipids are to be detected the eluting radioactivity may be monitored with either an on-line flow detector or with fraction collecting and scintillation counting. This alternative provides a very sensitive form of detection when compounds of high specific activity are present.

11.3.3. Sphingolipids

11.3.3.1. Long-chain bases and ceramides

Ceramides are amides of long-chain fatty acids with long-chain bases of which the commonest is sphingosine (*trans*-4-sphingenine). The long-chain bases may be released from glycosphingolipids by acid hydrolysis and then derivatised to form species which may be subjected to reversed phase HPLC. Determination of sphingoid bases in the picomolar range by pre-column derivatisation with the azo dye 4-dimethylaminoazobenzene-4-sulphonyl chloride (dabsyl-Cl) has recently been described (Rosenfelder et al., 1983). The stationary phase used was a Zorbax CN column (Du Pont Instruments) and the mobile phase used a gradient of acetonitrile–0.0175 M sodium acetate, pH 6.0 (90:60) which was increased to 90% acetonitrile. Using UV detection at 430 nm a detection limit of 5 pmol was reported and a wide range of sphingoid bases were separated. An alternative derivatisation procedure forms biphenylcarbonyl derivatives of the sphingoid bases after acid hydrolysis of the glycosphingolipids (Jungalwala et al., 1983; Kadowaki et al., 1983). Subsequent reversed phase chromatography used a stationary phase of Ultrasphere ODS (Altex) and mobile phases of either tetrahydrofuran–methanol–water (25:40:20) or methanol–water (94:6). Using UV detection at 280 nm a lower detection limit of less than 1 nmol was reported and a wide range of bases was resolved and detected including: C_{18}-5-hydroxy-sphingosine, C_{18}-erythro- and threo-sphingenine, sphinganine, C_{18}-5-O-methyl- and 3-O-methyl-sphingenine, C_{20}-sphingenine, sphingosylphosphatidylcholine and psychosine.

Ceramides have been resolved by both reversed phase HPLC and non-aqueous reversed phase HPLC on similar columns. Using an ODS stationary phase in combination with a mobile phase of chloroform–methanol, the benzoyl and *p*-nitrobenzoyl derivatives of both hydroxy and non-hydroxy ceramides from egg yolk and bovine brain were reported to be resolved (Do et al., 1981). The *p*-nitrobenzoyl derivatives were quantitated by UV detection at 260 nm and a lower limit of detection of 50 pmol was reported. An alternative derivatisation procedure involves the formation of perbenzoyl derivatives with chromatography on an ODS stationary phase in combination with a mobile phase containing varying proportions of methanol and acetonitrile (Yahara et al., 1980). This chromatographic system was shown to resolve the ceramides from the cerebrosides and sulphatides present in the peripheral nerves of adrenoleukodystrophy patients.

The benzoylated derivatives of both hydroxy and non-hydroxy ceramides have been resolved using non-aqueous reversed phase chromatography with a variety of mobile phases including either 0.05% methanol in n-pentane or mixtures of hexane–ethyl acetate (94:6, 95:5, 97:3) (Sugita et al., 1979; Iwamori and Moser, 1975; Iwamori et al., 1979).

11.3.3.2. Glycosylceramides

Glycosylceramides contain a ceramide unit which is linked via position 1 of the long-chain base to glucose or galactose or a polysaccharide unit. The term cerebrosides has been used to describe the latter. The most common HPLC techniques used for the separation and detection of the glycosylceramides involve pre-column derivatisation, usually to form the O-acylbenzoyl derivative with subsequent normal phase chromatography on a silica stationary phase and UV detection. The early pre-column derivatisation processes used benzoyl chloride as the derivatising agent (Evans and McCluer, 1972) although benzoyl anhydride has also been used for cerebrosides containing either non-hydroxy fatty acids (NFA-CR) or hydroxy fatty acids (HFA-CR) with subsequent chromatography on a silica stationary phase with a mobile phase of 0.13% methanol in pentane (McCluer and Evans, 1973). A more commonly used mobile phase for the resolution of the perbenzoyl derivatives of the glycosylceramides on silica stationary phases utilises a gradient of an increasing concentration of dioxane in hexane (Jungalwala et al., 1977; Ullman and McCluer, 1978; Lee et al., 1982). A gradient increasing from 1 to 23% dioxane in hexane has been suggested for the separation of the perbenzoyl derivatives of glycosylceramides containing less than six sugar residues, while a gradient increasing from 7 to 31% dioxane in hexane was suggested for the perbenzoyl derivatives of gangliosides

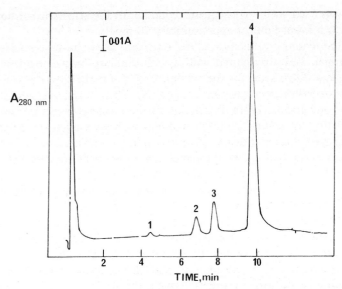

Fig. 11.3.3. HPLC separation of glycolipids from human erythrocytes. Chromatographic conditions: column, stainless steel tube (5000 × 2.1 mm) packed with a pellicular silica gel packing with an average particle size of 27 μm; mobile phase, 10 min
gradient of 2–17% aqueous ethyl acetate in hexane; flow rate, 2 ml/min; detection, UV
at 280 nm. The red blood cell neutral glycolipids were isolated from 5 ml of normal
blood. Peaks: 1, Glc-Cer; 2, Lac-Cer; 3, Gal-Lac-Cer; 4, globoside. Reproduced from
Ullman and McCluer (1977), with permission.

and glycosylceramides containing more than six sugar residues (Lee et
al., 1982). The quantitation of these derivatives used UV detection at
230 nm which gave a lower limit of detection of approximately 10
pmol.

An alternative mobile phase which has been used for the resolution
of perbenzoyl derivatives used ethyl acetate in hexane, either isocrati-
cally (McCluer and Evans, 1976) or as a gradient increasing from 2 to
17% ethyl acetate in hexane (Ullman and McCluer, 1977). This
chromatographic system resolved the perbenzoyl derivatives of neutral
mono-, di-, tri- and tetraglycosylceramides (Fig. 11.3.3). The UV
opacity of ethyl acetate precluded monitoring at 230 nm and so 280

nm was used with a consequent reduction in the sensitivity of detection to a lower limit of approximately 70 pmol.

Modifications to the procedures described above have been to use a Micropak NH_2-10 column with a mobile phase of cyclopentane–2-propanol (98/5 : 1.5) for the resolution of the perbenzoylated derivatives of NFA-glycosylceramide and NFA-galactosylceramide (McCluer and Evans, 1976). Formation of the *O*-acetyl-*N*-*p*-nitrobenzoyl derivatives of neutral glycosylceramides with subsequent chromatography on a silica stationary phase in combination with a mobile phase gradient system of 1–5% 2-propanol in hexane–dichloromethane (2 : 1) and UV detection at 254 nm has been reported (Suzuki et al., 1980).

The neutral glycosphingolipids derived from murine myeloma cells and T-lymphocytes have been separated and detected without prior derivatisation using a silica stationary phase and a gradient system using mobile phases containing varying proportions of methanol and water in chloroform (Kniep et al., 1983). Both UV detection at 230 nm and radioactive detection of the eluting peaks were used for quantitation. Clearly, dispensing with the derivatisation procedure accelerates the overall speed of analysis but this is made with some sacrifice in the sensitivity of detection.

Low pressure HPLC with a stationary phase of porous beads of silica gel (Iatro beads; Iatron Laboratories, Tokyo) has been used for the resolution of neutral glycosphingolipids using a gradient system with a mobile phase of 2-propanol–hexane–water in which the proportions of hexane and water were varied (Watanabe and Arao, 1981; Kannagi et al., 1982).

11.3.3.3. Gangliosides

Gangliosides are complex ceramide polyhexosides that contain one or more sialic acid group(s). The most sensitive HPLC technique for the resolution and quantitation of the gangliosides by HPLC utilises pre-column derivatisation to form perbenzoyl gangliosides with subsequent normal phase chromatography on silica (Bremer et al., 1979; Lee et al., 1982). The mobile phases in these separations utilised linear gradients of 7–23% dioxane in hexane and provided resolution of the major monosialogangliosides GM1, GM2, GM3, GM4 and also the polysialogangliosides GD3, GD1A and GD1B (Bremer et al., 1979). It should be noted, however, that the separation of the perbenzoyl derivatives of glycosphingolipids both with and without sialic acid residues has been demonstrated under the same chromatographic conditions (Lee et al., 1982). Each of these separations used UV detection at 230 nm and a lower limit of detection of 50 pmol was reported.

An HPLC system for the resolution and detection of non-derivatised gangliosides has been reported using a silica stationary phase with a mobile phase of chloroform–methanol–HCl (60 : 35 : 4) with a final HCl concentration of 0.01 M (Tjaden et al., 1977). This system was able to resolve GM3, GM2, GM1, GD1a, GD1b and GT1 and used a moving wire system with a flame ionisation detector for quantitation of the eluting peaks. Linearity of detection was reported over the range 2–200 μg. An alternative system not relying on derivatisation used a linear gradient of 2-propanol–hexane–water (55 : 42 : 3–55 : 25 : 20) with a silica stationary phase for the separation of mono-, di- and trisialogangliosides (Kundu and Scott, 1982). Quantitation of the eluting components used fraction collection and subsequent thin layer chromatographic analysis.

Reversed phase chromatography has been used for the separation of gangliosides using an ODS stationary phase with a mobile phase gradient consisting of varying mixtures of chloroform, methanol and water (Iwamori and Nagai, 1978). More recently, a procedure for the preparation of p-nitrobenzyloxyamine derivatives of brain gangliosides has been described (Traylor et al., 1983) which utilised subsequent gradient chromatography on an ODS stationary phase with a mobile phase of methanol–water (50 : 50–70 : 30). The peaks eluting from the column were detected at 254 nm although it was suggested that by monitoring fluorescence at 525 nm quantitation of less than 100 ng of ganglioside would be achievable.

11.3.4. Acylglycerols

Included in this class of compounds are the mono-, di- and tri-acylglycerols in which the hydroxyl groups of the glycerol molecule are esterified to one, two or three fatty acid molecules, respectively. The most popular chromatographic technique used for the resolution of individual acylglycerol species has been conventional reversed phase chromatography with some recent work on non-aqueous reversed phase systems. The resolution of mono-, di- and triacylglycerols from other lipid components has been achieved with normal phase chromatography using silica-based stationary phases and we shall initially consider these separations.

A good example of the all round utility of silica as a stationary phase for the separation of a complex mixture of neutral lipids is shown in Fig. 11.3.4, where each of the acylglycerol species was resolved from sterols, sterol esters, triterpene esters and free fatty acids (Hammond, 1981). The mobile phase used in the separation was a gradient system which was initially toluene–hexane (1 : 1) and finally toluene–ethyl acetate (3 : 1) plus 1.2% formic acid. The eluting peaks were monitored with a moving wire flame ionisation detector. A similar separation of the neutral lipids and free fatty acids present in liver using a silica stationary phase has been demonstrated (Greenspan and Schroeder, 1982). In this system a mobile phase of 2,2,4-trimethylpentane–tetrahydrofuran–formic acid (90 : 10 : 0.5) was used together with a refractive index detector and provided a lower limit of detection of ~ 0.5–1.0 μg per peak. These authors pointed out that the use of reversed phase techniques in such separations was unhelpful in that, by resolving the individual acylglycerols, it overcomplicated the resulting chromatogram. Similar separations using silica stationary phases but with alternative mobile phases have been used in the chromatographic analysis of serum lipids (Aitzetmuller and Koch, 1978), linseed oil and soybean oil (Plattner and Payne-Wahl, 1979) and multiple acylglycerols (Payne-Wahl et al., 1979).

Argentation HPLC using silver nitrate-loaded silica stationary phases has been used for the analysis of the positional isomers of triglycerides including 2-unsaturated-1,3-disaturated (SUS), 1-unsaturated-2,3-disaturated (USS), fully saturated (SSS) and fully unsaturated (UUU) isomers (Smith et al., 1980; Hammond, 1981). In these separations the silica was loaded with 10% silver nitrate and the mobile phases used were either benzene (Smith et al., 1980) or toluene (Hammond, 1981). The resolution of the lipid species was found to

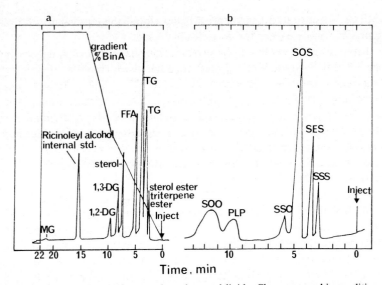

Fig. 11.3.4. (a) Silica HPLC separation of neutral lipids. Chromatographic conditions: column, Lichrosorb Si60 (5 μm) (100×4 mm I.D.); mobile phase, toluene–hexane (1 : 1) (Solvent A), toluene–ethyl acetate (3 : 1) plus 1.2% formic acid (solvent B); flow rate 1.5 ml/min; detector, Pye LCM2 transport detector. Peaks: MG, monoglycerides; DG, diglyceride; FFA, free fatty acid; TG, triglyceride. (b) Separation of triglycerides by AgNO₂ HPLC. Chromatographic conditions: column, Partisil (10 μm) (250×4 mm I.D.) saturated with 10% AgNO₃; mobile phase, toluene; flow rate, 1 ml/min; temperature, 6° C; detector, refractive index. Peaks: SSS, trisaturated; SES, 2-elaido-1,3-distearin; SOS, 2-monounsaturated; SSO, 3-monounsaturated; PLP, 2-linoleoyl-1,3-dipalmitoylglycerol; SOO, 2,3-diunsaturated. Reproduced from Hammond (1981), with permission.

improve as column temperature was reduced and in these separations the temperature was maintained at 6.0–6.8° C.

Some of the earliest bonded phase chromatographic separations of triacylglycerols used a stationary phase of hydroxyalkoxypropyl Sephadex under low pressure conditions in combination with mobile phases of either acetone–water–heptane (87 : 13 : 10) plus 1% pyridine to reduce acid hydrolysis (Curstedt and Sjovall, 1974) or a gradient system of 2-propanol–chloroform–hexane–water (115 : 2 : 15 : 35) and heptane–acetone–water (5 : 15 : 1) (Lindqvist et al., 1974). In these separations good resolution of C_9 to C_{54} triglycerides was reported with the unsaturated triglycerides being eluted the equivalent of 1.42 methylene units earlier for each double bond.

More recently, reversed phase separations have concentrated on the use of ODS stationary phases together with mobile phases of acetone–acetonitrile (Plattner et al., 1977, 1978; Bezard and Ouedraogo, 1980; El-Hamdy and Perkins, 1981; Tsimidou and Macrae, 1984). In these systems the separation of the individual triglycerides was on the basis of chain length of the individual esterified fatty acids and the degree of unsaturation. As the length of the fatty acid chain was increased the retention time of the triglycerides was increased, while the introduction of a double bond reduced the retention time to that of a triglyceride with approximately two less methylene units. Similarly, triglycerides containing hydroxy fatty acids such as ricinoleic acid were eluted slightly before normal triglycerides (Plattner et al., 1977) and the presence of epoxy groups in fatty acids also greatly reduced the retention time (Plattner et al., 1978). The characteristics of the stationary phase were also shown to be important, with better resolution of the triglycerides being obtained from columns with either an increased carbon loading or reduced particle size (El-Hamdy and Perkins, 1981). A practical consideration for the chromatographer planning to use HPLC for the analysis of triglycerides is that some care must be taken in the selection of the solvent in which the sample is dissolved since it has an unusually large effect on column efficiency. Ideally, the mobile phase should be used as the reconstituting solvent after extraction, but where this is impractical because of solubility problems the best solvent to use is acetone, while the use of chloroform, diethyl ether and tetrahydrofuran should be strictly avoided due to their detrimental effects on column efficiency (Tsimidou and Macrae, 1984).

A series of alternative reversed phase systems have been used for the separation of individual triglyceride species on the basis of their chain length and degree of unsaturation. For example, the first conventional reversed phase HPLC separation utilised a mobile phase of 60% aqueous methanol (Pei et al., 1975). It should be noted that the substitution of a methanol–acetone mobile phase for an acetonitrile–acetone phase causes small changes in retention time, with the introduction of one double bond in a fatty acid being equivalent to 2.2 carbon atoms in the first mobile phase and equivalent to 2.4 carbon atoms in the second mobile phase (Herslof et al., 1979). A mobile phase of acetonitrile–tetrahydrofuran–hexane (224 : 123.2 : 39.6, w/w/w) has been demonstrated to be useful in the resolution of critical pairs of triglycerides (Jensen, 1981). Similar chromatographic separations of tributyrin, tricaprylin, trilaurin and tripalmitin have been reported with mobile phases of either methylene chloride–acetonitrile (40 : 60) or tetrahydrofuran–acetonitrile (40 : 60) (Parris, 1978). The use of propionitrile–acetonitrile mixtures in a gradient HPLC system has recently been demonstrated to provide good resolution of individual molecular species of triglycerides and it was suggested that the precise ratio of the two solvents was less critical than in acetone–acetonitrile mixtures (Marai et al., 1983; Myher et al., 1984). However, as a word of caution, the propionitrile should be distilled over phosphorous pentoxide prior to use to remove alkaline impurities that would otherwise accumulate in the solvent. Finally, a rather unusual mobile phase and stationary phase combination has been reported using an ODS column in combination with a mobile phase of methanol–2-propanol (3 : 1) plus 0.05 N silver chlorate (Vonach and Schomburg, 1978). The addition of silver ions to the stationary phase was reported to improve the resolution of trilinolenin, triolein and trielaidin, with separations being effected on the basis of chain length and degree of unsaturation.

The most popular methods of monitoring the eluting peaks in the reversed phase separations described above have been to use refractive index detectors and flame ionisation detectors. While the latter are no longer commercially available, technological advances in refractive index detectors are constantly improving their lower limits of detection and for triglycerides this is in the range of 0.5–1.0 μg. A non-aqueous reversed phase system has been used in combination with an infra-red detector monitoring at a wavelength of 1700 cm^{-1} (Parris, 1978). More recently, triglycerides have been analysed using a quadrupole mass spectrometer with the column eluent being admitted via a direct liquid interface (Marai et al., 1983). The limit of detection for a sample using this technique was approximately 50 ng, but at least 100 times more material was required for column injection as only 1% of the column eluent was injected into the mass spectrometer.

Finally, gel permeation chromatography has been utilised for the resolution of acylglycerols and molecular species of triglycerides. Styragel columns eluted with tetrahydrofuran have been used for the resolution of free glycerol and mono-, di- and triacylglycerols with the eluent being monitored with a refractive index detector (Aitzetmuller, 1983). The distribution of triglycerides in serum lipoproteins has also been analysed using a TSK GEL permeation column (Toyosoda, Japan) which was eluted with 0.15 M sodium chloride (Hara et al., 1982). The eluting triglycerides were subjected to a series of post-column enzymatic derivatisation reactions with the final step converting the compound N-ethyl-N-(3-methylphenyl)-N-acetylethylenediamine to a quinone diamine dye which absorbed strongly at 550 nm. The lower limit of detection using this derivatisation procedure was reported to be approximately 1 μg per eluted peak (Hara et al., 1982).

Carbohydrates

11.4.1. Introduction

The term carbohydrate includes a wide range of molecules which in many cases are quite complex structures. In chemical terms a carbohydrate is either a polyhydroxy aldehyde, a polyhydroxy ketone or, alternatively, it is a compound that can be hydrolysed to such a structure. The smallest unit that cannot be hydrolysed any further is called a monosaccharide. Glucose (Fig. 11.4.1) is the most abundant monosaccharide known and is by far the most important. Other examples include fructose and mannose. A carbohydrate which has been hydrolysed to two monosaccharide units is callled a disaccharide and the best known example is sucrose, which is composed of a unit of glucose covalently linked to a unit of fructose. The nomenclature system continues in a logical fashion until one can refer to an oligosaccharide as being a chain composed of several monosaccharide units. Depending upon whether a monosaccharide is an aldehyde or a

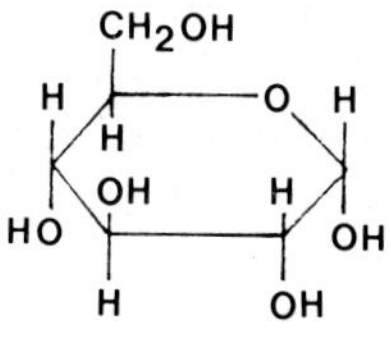

Fig. 11.4.1. Chemical structure of glucose.

ketone the molecule can be further classified as either an aldose or a ketose. Thus according to the composition of the carbon chain, carbohydrates can be referred to as an aldohexose such as glucose or a ketohexose such as fructose. Carbohydrates that act as reducing agents are classified as reducing sugars.

Carbohydrates are contained in food (e.g. they are present in grain and vegetables) and provide our clothing (cotton, linen, rayon, cellulose) and shelter (wood). The biosynthesis of carbohydrates begins with the production of glucose from carbon dioxide and water during photosynthesis in plant cells. The glucose can then be converted into either cellulose, which provides the supporting framework of the plant, or starch, which provides a food store. When starch (and cellulose in some species) is digested by animals it is metabolised into glucose and used directly as a source of energy (glycolysis) or stored as a source of energy in the form of glycogen (glycogenesis). Smaller quantities of glucose are converted into fats or used to make up glycoproteins.

HPLC has only recently been employed for the analysis and purification of mono- and oligosaccharides and consequently is still in the process of development and improvement. The classical method of analysis of carbohydrates, descending paper chromatography, was used routinely for many years. The disadvantage of this procedure was the time involved in the development of the chromatogram and the requirement for repeated chromatographic runs to ensure a sufficient concentration of components to allow successful detection. An alternative method of analysis for saccharides has been the use of gas chromatography, although the requirement for volatile components has necessitated pre-column derivatisation with a suitable reagent. Such derivatives have included trimethylsilyl ethers (Wood and Siddiqui, 1971), butyl boronic acid (Eisenberg, 1971), substituted benzene boronic acid and alditol acetates (Crowell and Burnett, 1967). The extended sample preparation time, potential instability and potential for incomplete reactions of this technique make it unsuitable for routine or reproducible analyses, although it has been used for the identification of anomeric forms.

HPLC has been used extensively for the separation of mono- and oligosaccharides and its advantages have been well documented (Thiem et al., 1978; Van Olst and Joosten, 1979). HPLC offers accurate quantitation and rapid analyses for underivatised mono- and oligosaccharides (Verhaar and Kuster, 1981b). However, an inherent problem in HPLC analysis is that saccharides do not contain a suitable chromophore for their detection by standard UV systems and pre-column derivatisation is consequently often preferred (see below). Alternatively, other detection methods, such as those monitoring refractive index, can be used (Palmer, 1975; White et al., 1981).

11.4.2. Modes of separation

Unmodified silica is too polar to be effective in resolving carbohydrates except where initial pre-column derivatisation has been employed. Separations of saccharides by HPLC have used a variety of stationary phases including:
(1) Amine or nitrile bonded phases (D'Amboise et al., 1980; Macrae and Dick, 1981; Boersma et al., 1981; Lee et al., 1981).
(2) Silica modified in situ with amines (Wheals and White, 1979; White et al., 1981; Aitzetmuller, 1978).
(3) Cation-exchange resins (Verhaar and Kuster, 1981a).
   The mechanism by which saccharides are retained on amine columns has been reviewed elsewhere (Verhaar and Kuster, 1981b, 1982) and will not be discussed in detail in this section.

11.4.3. Sample preparation

In most situations simple aqueous extraction at elevated temperature is sufficient to extract the low molecular weight saccharides from food products, although organic solvent extraction and pulp blending can be employed in more difficult situations (Wilson et al., 1982a). Often these extracts are acidic and the use of calcium carbonate to neutralise acidity (with a trace of octanol to reduce frothing) is recommended. Extracts can be clarified with neutral lead acetate followed by deionisation and filtration (Wight and van Niekerk, 1983a, 1983b). Subsequent dilution with acetonitrile and sample filtration or centrifugation is recommended prior to HPLC analysis. The use of a guard column is advised to protect the analytical column from potential contaminants such as proteins.

The solubility of oligosaccharides in acetonitrile decreases with increasing molecular weight and typically one would use 25% water in acetonitrile to elute monosaccharides and up to 45% water in acetonitrile to elute oligosaccharides up to 15 units long from a chemically modified alkylamine column.

Generally, soft drinks, wines and fruit juices (Ondrus et al., 1983) can be directly injected after an initial filtration step. Non-cellulose cell wall polysaccharides can be released from cells by acid hydrolysis (2 M trifluoroacetic acid, 1 h, 120° C) followed by either deionisation on a mixed bed resin or elution through a Sep-Pak cartridge (Barton et al., 1982). Complex oligosaccharides from urine can be separated from salts and metabolites by adsorption onto activated charcoal (Warren et al., 1983) with subsequent elution using aqueous ethanol to recover the oligosaccharides. Glycoproteins contain complex oligosaccharides covalently linked to the protein portion, usually through an asparagine residue. HPLC analysis of these oligosaccharides can be performed by prior release of the oligosaccharide by controlled hydrazinolysis (Endo et al., 1982).

More extensive clean-up procedures for saccharide samples utilise passage through a micro-column such as a Sep-Pak C_{18} cartridge (Waters Assoc.) (Richmond et al., 1981). This procedure removes contaminants such as lipids and increases column lifetime. It is also important that samples should be properly deionised prior to analysis, although problems have been experienced using anion-exchange resins for this purpose particularly when in the hydroxyl ion form (Baust et al., 1982). These difficulties can be overcome by employing the formate ion form followed by freeze drying to remove the formic acid.

11.4.4. Pre-column derivatisation

One of the main problems in the analysis of saccharides is in their detection since they have minimal UV absorption at wavelengths above 200 nm; consequently, they are often detected by refractive index (RI) monitoring. For many molecules the limited sensitivity available with RI detectors is both insufficient and impractical and therefore derivatives with a higher absorption at 254 nm have been prepared.

Pre-column derivatisation of saccharides should aim to generate a derivative that: (*a*) is easy to prepare in high yield in a reproducible manner; (*b*) absorbs strongly at a suitable wavelength; (*c*) is compatible with preferred solvent systems; *d*) allows easy recovery of the precursor and (*e*) is suitable for further analysis, e.g. by NMR or by mass spectrometry (White et al., 1983). In addition the chemical reactivity and stability must be established. A number of derivatives of saccharides have been described, including acetates (Wells and Lester, 1979), benzoates, 4-nitrobenzoates (Nachtmann and Budna, 1977), benzyloxime perbenzoates (Thompson, 1978), other substituted benzyloximes (Chen and McGinnis, 1983), phenyldimethylsilyl derivatives (White et al., 1983) and alkylated derivatives (McGinnis and Fang, 1978). Reaction with *p*-anisidine can be employed which, in the absence of acid catalysis, will react specifically with aldoses as opposed to ketoses (Batley et al., 1982).

For the analysis of complex oligosaccharides which can often be present in quite low concentrations, radioactive labelling with tritiated sodium borohydride has been used and allows more accurate and sensitive quantification (Warner et al., 1983; Mellis and Baenziger, 1983).

In general, derivatisation procedures reduce the polarity of compounds and facilitate their separation on underivatised silica using organic mobile phases (White et al., 1983). Generally, anomeric forms can be separated after acetylation or benzoylation although the increased number of peaks due to these anomeric forms necessitates longer elution times and gives more complex chromatograms. Single peaks are obtained when derivatisation causes ring opening; this is found particularly with the oxime reagents (Thompson, 1978).

11.4.5. Separation of simple saccharides

The term simple saccharide is used in the present context to refer to molecules consisting of between 1 and 6 saccharide units and thus includes such compounds as fructose, glucose, lactose, maltose, sucrose, galactose and arabinose.

11.4.5.1. Chemically modified alkylamine columns

Chemically modified alkylamine columns are the most widely used stationary phase for saccharide separations and are generally eluted with mobile phases consisting of acetonitrile–water mixtures. This mixture has been found to be far superior to others such as methanol–water (Rabel et al., 1976) and ethyl acetate–ethanol–water (Binder, 1980). It has been observed that as the water content of the mobile phase is increased the retention of saccharides is decreased. The columns are usually stable for periods of up to 3 months but

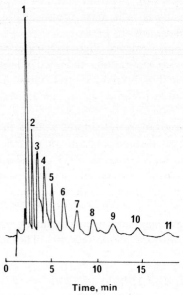

Time, min

Fig. 11.4.2. Analysis of chicory root extract. Numbers refer to length of saccharide chains. Chromatographic conditions: column, Spherisorb S5 NH$_2$ (250×4.6 mm); mobile phase, acetonitrile–water (60:40); flow rate, 2.0 ml/min; detection, refractive index. Reproduced from Macrae (1982), with permission.

eventually amine groups are lost and retention times are decreased. The ability of primary amines to form Schiff bases with reducing saccharides (Ellis and Honeyman, 1955) is the main limitation of these columns, but for the common mono- and disaccharides this is not a problem (Abbott, 1980). Secondary amines are less reactive towards reducing saccharides and it is therefore recommended that stationary phases modified with secondary amines (e.g. Partisil 10-PAC) are used for the resolution of reducing saccharides.

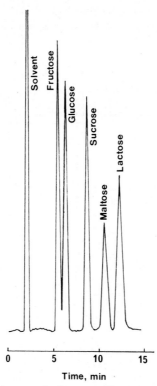

Fig. 11.4.3. Analysis of common food sugars. Chromatographic conditions: column, Spherisorb S5 NH$_2$ (250×4.6 mm); mobile phase, acetonitrile–water (75:25); flow rate, 2.0 ml/min; detection, refractive index. Reproduced from Macrae (1982), with permission.

An analysis of chicory root extract clearly demonstrates the useful-ness of chemically modified alkylamine columns for the separation of saccharides according to their degree of polymerisation (Fig. 11.4.2). By changing the ratio of acetonitrile and water in the solvent from 60 : 40 to 75 : 25 these columns show considerable versatility in the separation of individual food saccharides (Fig. 11.4.3). Using a very similar column and eluent, monosaccharides from storage cell walls have been successfully separated and therefore this type of analysis can be used to monitor relative nutritional value; it was also noted that the often problematical separation of galactose from glucose was achieved with this system (Barton et al., 1982). The combination of chemical modification (substituted oximes with UV detection) and a

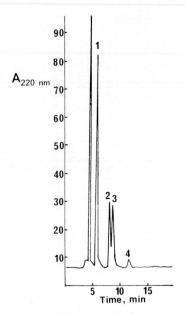

Fig. 11.4.4. Separation of oxime derivatives of saccharides. Chromatographic condi-tions: column, polar-bonded amino column (Lichrosorb NH$_2$); mobile phase, acetonitrile–water (80 : 20); detection, UV at 220 nm. Peaks are O-methyloximes of: 1, D-glucose; 2, D-maltose; 3, D-cellobiose; 4, D-maltotriose. Reproduced from Chen and McGinnis (1983), with permission.

chemically bonded amine column has been used to effect the separation of both monosaccharides and small oligosaccharides (Fig. 11.4.4).

A dual column system has been described in which a complex mixture of saccharides from cell wall hydrolysates were separated on a chemically bonded amine column (Blaschek, 1983). The saccharides in the unresolved peaks were then further resolved by using cation-exchange chromatography which facilitated a baseline separation of all the carbohydrate components present in the extract.

11.4.5.2. Physically modified amine columns

In this technique a small quantity of an alkylamine is added to the mobile phase so that the silica acquires an amine coating by simple adsorption (Wheals and White, 1979; White et al., 1981). A number of different amine modifiers have been investigated for this technique including a polyfunctional amine (Aitzetmuller and Koch, 1978), diamines and monoamines (Wheals and White, 1979) with the final choice of amine being dependent upon the particular separation although the most widely used modifier is tetraethylenepentamine. Typical concentrations of tetraethylenepentamine used in the mobile phase are 0.01%, although initial pre-equilibration of the column with 0.1% has been used to speed up the process (Aitzetmueller, 1980). The resulting high pH of the eluent (pH 9.2) can cause dissolution of the silica and a guard column before the injector is recommended. Alternatively a radial compression system can be employed to fill the voids created by the loss of silica (Buckee and Long, 1982; Baust et al., 1983). The optimisation of such systems with regard to mobile phase, pH, solvent, amine modifier concentration and flow rate has been described (Hendrix et al., 1981) and excellent control over resolution can be obtained by varying one or more of these parameters.

A comprehensive survey of 63 low molecular weight saccharides compares the potential of a radially compressed physically modified amine column with that of a cation-exchange column and contains a listing of all the capacity factors (Baust et al., 1983).

The separation of food and plant saccharides on a silica column using a polyfunctional amine dissolved in the eluent has been demon-

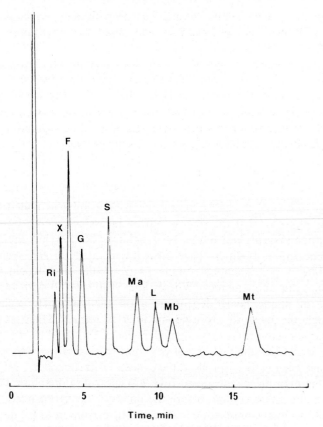

Time, min

Fig. 11.4.5. Separation of simple saccharides. Chromatographic conditions: column, LiChrosorb Si 60 (5 μm) (250×4.0 mm); mobile phase, acetonitrile–water (75:25) containing 0.01% amine modifier; flow rate, 3.0 ml/min; detection, refractive index. Peaks: Ri, ribose; X, xylose; F, fructose; G, glucose; S, sucrose; Ma, maltose; L, lactose; Mb, melibiose; Mt, maltotriose. Reproduced from Wight and Van Niekerk (1983b), with permission.

strated to give excellent resolution (Fig. 11.4.5). Undesirable chemical reactions between the amine modifier and samples were negligible under the experimental conditions. It has been shown that the k' values for the common saccharides on this column were similar to those obtained from a chemically bonded amine column.

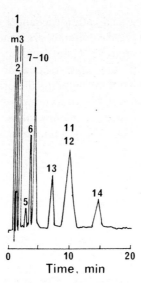

Fig. 11.4.6. Analysis of a carbohydrate mixture on an amine modified silica column. Chromatographic conditions: column, Waters Sugar-Pak I; mobile phase, 10^{-4} M calcium acetate; flow rate, 0.5 ml/min; temperature, $90\,^{\circ}$C. Peaks: 1, ethanol; 2, ethylene glycol; 3, glycerol; 5, xylose; 6, fructose; 7, galactose; 8, glucose; 9, mannitol; 10, sorbitol; 11, lactose; 12, maltose; 13, sucrose; 14, melezitose; m, artifact due to mobile phase. Reproduced from Baust et al. (1983), with permission.

An example of the separation of 14 saccharides by amine modified silica is shown in Fig. 11.4.6. The individual components were separated into eight groups in an order approximating that of increasing molecular weight, whereas using ion-exchange chromatography only 6 groups could be obtained in roughly the reverse order. The authors conclude that the silica system offers long life, low cost and high resolution with a wide sample range although ion-exchange offers greater sensitivity through smaller band broadening effects.

11.4.5.3. Cation-exchange columns

The new resin-based specialised carbohydrate columns utilise an 8% cross-linked styrene–divinylbenzene copolymer which is functionalised to produce a strong acid cation-exchanger (Brando et al., 1980; Richmond et al., 1981). These columns are generally operated at an elevated temperature of up to 85°C but lack the ability to successfully resolve components with the same degree of polymerisation, e.g. maltose and sucrose. Using these columns the preferred counter-ion is calcium and they can be initially conditioned with 10^{-3} M calcium-acetate followed by the required mobile phase containing 10^{-4} M calcium acetate (Baust et al., 1983). Approximately 12 h are required

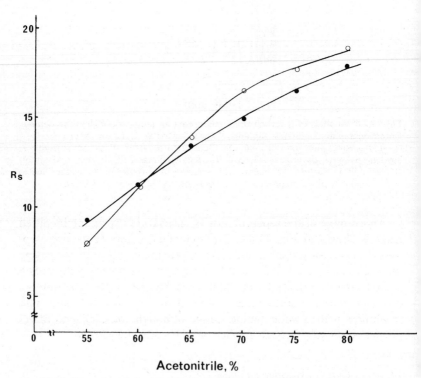

Fig. 11.4.7. Dependence of resolution (R_s) of glucose and galactose (○) and mannitol and glucitol (●) on acetonitrile concentration. Chromatographic conditions: column, 10% cross-linked sulphonated HC095AA PS-DVB resin; mobile phase, acetonitrile–water of varying concentrations; temperature, ambient. Reproduced from Kawamoto and Okada (1983), with permission.

for pre-equilibration. Sodium counter-ions offer a different selectivity which can be valuable in certain situations (Kawamoto and Okada, 1983).

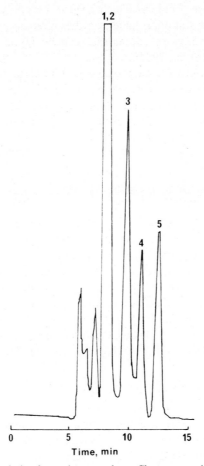

Fig. 11.4.8. HPLC analysis of strawberry yoghurt. Chromatographic conditions: column, Bio-Rad HPX-87 carbohydrate column; mobile phase, water; flow rate, 0.6 ml/min; temperature, 80°C. Peaks: 1, sucrose; 2, lactose, 3, glucose; 4, galactose; 5, fructose. Reproduced from Richmond et al. (1982), with permission.

Separation of difficult pairs of saccharides such as glucose and galactose, and mannitol and glucitol is strongly dependent upon the acetonitrile concentration (Fig. 11.4.7). The system shown demonstrates good separations for most monosaccharides and for mixed di- and trisaccharides. Nevertheless, certain combinations of saccharides still pose problems which need to be resolved by alternative mobile phases or stationary phases. Other references to the separation of standard mixtures of saccharides can be found in the literature (Vidal-Valverde et al., 1982).

The application of cation-exchange HPLC to the separation of saccharides in food is illustrated in Fig. 11.4.8 with the separation of strawberry yoghurt saccharides (Richmond et al., 1982). Cation-exchange HPLC has also been used for the separation of saccharides from molasses (Abeydeera, 1983) and from wood pulp (Wentz et al., 1982).

11.4.5.4. Reversed phase columns

Reversed phase HPLC is only rarely used for the chromatographic resolution of saccharides; however, in instances where only occasional separations are required the wide availability of reversed phase columns in many laboratories saves the purchase of a specialised carbohydrate column. Reversed phase HPLC has been used primarily for the separation of oligosaccharides and separation is dependent on the degree of polymerisation (Cheetham et al., 1981). Organic phase modifiers such as n-alkylamines can be used to provide increased capacity and selectivity for saccharides (Lochmuller and Hill, 1983).

11.4.6. Separation of oligosaccharides

HPLC analysis of oligosaccharides from animal tissues or body fluids has been used to determine their molecular weight distribution. The majority of complex oligosaccharide separations has been carried out using chemically bonded amine columns which separate molecules on the basis of chain length, although reversed phase HPLC has been used to separate human milk oligosaccharides (Dua and Bush, 1983).

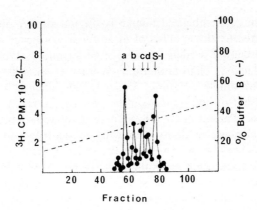

Fig. 11.4.9. Separation of monosialylated complex oligosaccharides. Chromatographic conditions: column, MicroPak AX-5 (Varian Assoc.); mobile phase, linear gradient of acetonitrile–water (80:20) containing 3% triethylammonium acetate (pH 5.5) to 100% water containing 3% triethylammonium acetate (pH 5.5), initial concentration of Buffer B was 15% and was increased at the rate of 0.75%/min; flow rate, 1 ml/min. Peak S-I is a dibranched, complex oligosaccharide bearing one sialic acid residue. Peaks b, c and d were prepared from S-I by sequential digestion with jack bean beta-galactosidase (d), beta-hexosaminidase (c) and alpha-mannosidase (b). Peak (a) is from a patient with sialidosis. Reproduced from Mellis and Baenziger (1983), with permission.

The reversed phase system was a useful complement to the amine columns since it was sensitive to stereochemical differences in molecules. Many oligosaccharides which are present as part of glycoproteins are negatively charged because of the presence of sialic acid, phosphate or sulphate residues. Anion-exchange HPLC is capable of separating such glycoproteins according to differences in charge (Baenziger and Natowicz, 1981). Amine bonded columns eluted with water–acetonitrile containing 3% triethylammonium acetate (pH 5.5) can be used to separate glycoproteins according to their net carbohydrate content. The added salt effectively suppressed the ionic effects (Figure 11.4.9). Similarly, complex sialyloligosaccharides from both monkey and human have been separated using this method (Lamblin et al., 1983) and separations of the isomeric oligosaccharides present within the complex chains of glycoproteins (Bergh et al., 1983) and hyaluronic acid also have been recently described (Nebinger et al., 1983).

The diagnosis of inherited human lysosomal storage disorders has been facilitated by the analysis of urine for the presence of oligosaccharides containing a high percentage of mannose (Warren et al., 1983) or galactosyl oligosaccharides (Warner et al., 1983) on chemically bonded amine columns. Similarly the analysis of oligosaccharides present in hen ovomucoid has been undertaken after initial hydrazinolysis (Parente et al., 1982). These analyses have been achieved in the normal phase mode by employing acetonitrile–water mixtures as eluents. Generally an increase in the water content of the mobile phase to 45% allows elution of oligosaccharides of up to 15 units long at room temperature. An increase in the column temperature reduces retention times and enhances elution of longer oligosaccharides. To elute much larger oligosaccharides (e.g. starch) aqueous size exclusion chromatography is necessary.

11.4.7. Detection

The most common HPLC detection method is to monitor UV absorption. Unfortunately, saccharides do not have a suitable chromophore to permit detection in the wavelength range 230–300 nm. Although they can be detected in the region 185–195 nm, the requirement for very high quality solvents in this range of wavelength precludes the general use of this technique. However, it should be noted that for both glucose and fructose, the sensitivity of detection using UV monitoring may be better than by using refractive index detection (Binder, 1980).

Chemical modification of saccharides is an ideal tool for shifting their absorption maximum into a range which facilitates classical UV detection but suffers from inherent problems (Section 11.4.4) which precludes the popular use of this methodology. Detection of changes in refractive index is the preferred method of detection of the saccharides and detection limits are in the range 10–40 μg and depend upon the specific system employed. More recently mass detectors have been described which can be used with gradient elution and which achieve a 10 times greater sensitivity than refractive index detectors (Macrae and Dick, 1981). Less common techniques for the detection of saccharides include scintillation counting (Mellis and Baenziger, 1981) and electrochemical detection (Buchberger et al., 1983).

11.4.8. Post-column derivatisation

A popular alternative for the detection of saccharides is to use post-column derivatisation. The most popular method is to derivatise the saccharide with tetrazolium blue which allows absorption to be measured at 530 nm (D'Amboise et al., 1980). This reaction requires reducing saccharides and is therefore applicable to many components but will exclude, for example, sucrose and inulin. In such instances pre-column treatment with beta-fructosidase to convert sucrose to glucose (Wight and van Niekerk, 1983b) or mild acid hydrolysis to convert inulin to fructose and glucose is necessary (Wight and van Niekerk, 1983). Alternative post-column derivatisation reagents include production of a saccharide–cuprammonium complex which can be detected at 295 nm (Grimble et al., 1983), reaction with periodate and detection by loss of absorbance at 260 nm (Nordin, 1983), reaction with 2-tert-butylanthraquinone and detection by photoreduction fluorescence (Gandelman and Birks, 1983) and reaction with cerium(IV) and detection by fluorescence measurement (Mrochek et al., 1975).

Prostaglandins, leukotrienes and hydroxyeicosatetraenoic acids

Since the isolation and identification of the first prostaglandins between 1957 and 1964 by Bergstrom and coworkers there has been a vast expansion in research associated with the role of these compounds in the aetiology of a number of pathological states. All of the major prostaglandins, leukotrienes and hydroxyeicosatetraenoic acids (HETEs) are derived from arachidonic acid and have been identified as important mediators of inflammatory and allergic reactions. As each of these groups of compounds is of major interest in many areas of biological research a number of different assay systems have been devised to allow quantitation including: bioassay, gas chromatography–mass spectrometry, radioimmunoassay (RIA), thin layer chromatography (TLC) and, more recently, HPLC. Bioassay provides a sensitive assay system but tissues need to be selected for each specific prostaglandin or leukotriene. This prerequisite may not always be possible but even where specificity can be established the assay is time-consuming. Gas chromatography followed by mass spectrometry is a very sensitive technique and can yield very accurate structural information, but unfortunately sample preparation is extremely time-consuming, and clearly the equipment required is very expensive.

Developments in RIA over recent years have provided great improvements in both the sensitivity and specificity of the assay techniques, and this is probably the most commonly employed assay method in this field; however, a given RIA can only be used to quantitate one compound at a time so that if a complete prostaglandin and leukotriene profile is required the technique is excessively time-consuming. It is because of the drawbacks of these techniques that TLC and HPLC methodologies have been developed, with recent research tending to concentrate on HPLC because of its numerous advantages over TLC. It should also be emphasised that HPLC is often used for sample preparation prior to the use of RIA for quantitative determinations, or of mass spectrometry for structural analyses.

11.5.1. HPLC of prostaglandins

Several chromatographic modes have been used for HPLC of prostaglandins including normal phase, reversed phase and silver ion-loaded cation-exchange chromatography. A number of HPLC separations using normal phase stationary phases have been reported, both with and without prior derivatisation procedures to facilitate subsequent detection. The most popular stationary phase used has been silica in combination with a number of different mobile phases with both isocratic and gradient modes. For example, one of the earliest recorded separations of prostaglandins utilised a mobile phase of ethyl acetate–acetic acid (80 : 20) to separate 15-*epi*-PGF$_{2\alpha}$ from PGF$_{2\alpha}$ (Weinshenker and Longwell, 1972). Other workers have used a mobile phase of chloroform–ethyl acetate–formic acid (84 : 15.7 : 0.3) to resolve PGB$_2$ (Dunham and Anders, 1973). Prior derivatisation of prostaglandins with 4-bromo-7-methoxycoumarin has been reported to allow resolution and quantitation of PGF$_{2\alpha}$, 6-keto-PGF$_{1\alpha}$, thromboxane B$_2$ (TxB$_2$), PGE$_2$ and PGD$_2$ using a mobile phase of chloroform–2,2,4-trimethylpentane–methanol (Turk et al., 1978).

A typical chromatogram produced in a normal phase separation is shown in Fig. 11.5.1. where a gradient separation was used with an initial mobile phase of chloroform and a final mobile phase of chloroform–methanol–acetic acid (93.4 : 6 : 0.6). This chromatogram demonstrates the major limitations associated with normal phase HPLC of prostaglandins where co-chromatography is often a problem

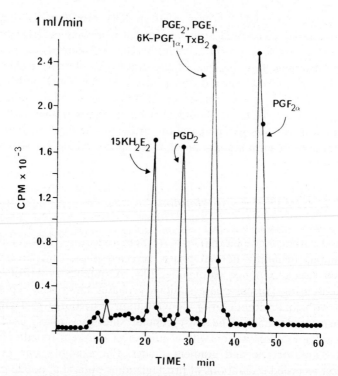

Fig. 11.5.1. HPLC separation of tritiated prostaglandins. Chromatographic conditions: column, μPorasil (10 μm) (Waters Assoc.); mobile phase, a 60 min linear gradient from chloroform (100%) to chloroform–methanol–acetic acid (93.4:6:0.6) was used; flow rate, 1 ml/min; detection, one-minute fractions were collected and assayed for radioactivity. The peaks shown represent 1–2 μg or 0.05–0.1 μCi of tritiated compound. 15KH$_2$E$_2$, 15-keto-13,14-dihydro PGE$_2$. Reproduced from Whorton et al. (1979), with permission.

especially with TxB$_2$ and 6-keto-PGF$_{1\alpha}$, both of which are important prostanoid derivatives. It is because of these problems of selectivity that normal phase chromatography has been largely superceded by the reversed phase mode.

Reversed phase is the most popular chromatographic mode used for the separation of prostaglandins. The most commonly used stationary and mobile phases are ODS in combination with a mobile phase of acetonitrile and water adjusted to pH 2–3 by the addition of low concentrations of either acetic or phosphoric acid. To elicit a full separation of metabolites derived from arachidonic acid in the reversed phase mode, it is necessary to use a gradient system in which the acetonitrile concentration is increased from 20–30% to 95–100%. Generally, acetonitrile is used as the organic modifier since it provides several distinct advantages over methanol; firstly, acetonitrile possesses a relatively high optical transparency at 192 nm, a wavelength commonly used for the spectrophotometric detection of prostaglandins, while methanol has an optical cut-off wavelength of approximately 210 nm. Secondly, phosphate salts are more soluble in acetonitrile than in methanol, thereby allowing the use of higher salt concentrations in the mobile phase. Finally, prostacyclin tends to undergo isomer formation in the presence of methanol (Wynalda et al., 1979).

An example of an isocratic reversed phase separation of the major prostaglandin species is shown in Fig. 11.5.2. It is apparent that a baseline separation of TxB_2 and $PGF_{2\alpha}$ is not achieved and that PGE_2 and PGD_2 are not resolved. However, by the introduction of a gentle gradient of acetonitrile, a baseline separation can be achieved (Van Rollins et al., 1982; Peters et al., 1983). It is often noticeable that even under gradient conditions peak broadening of TxB_2 occurs on ODS stationary phases under the acidic mobile phase conditions generally used in separations. The cause of the broad asymmetric peaks of TxB_2 is thought to be the existence of an equilibrium between an open and a closed hemiacetal ring in the structure of TxB_2 at acidic pH (Hamberg et al., 1975). The equilibrium between the two isomeric forms could be shifted by altering the mobile phase to a neutral pH but this, however, can result in peak tailing on silica stationary phases. A solution to this problem is to use an alternative non-silica-based stationary phase (PRP-1; Hamilton Company) in conjunction with a mobile phase gradient of 21% to 100% acetonitrile plus 1.2% (v/v) triethylamine in distilled water adjusted to pH 7.5 with acetic acid. Using these conditions, a sharp peak was obtained for the elution of TxB_2, together with a good separation from 6-keto-$PGF_{1\alpha}$, $PGF_{2\alpha}$ and PGE_2 (Moonen et al., 1983).

The use of silver ions to enhance the separation of prostaglandins

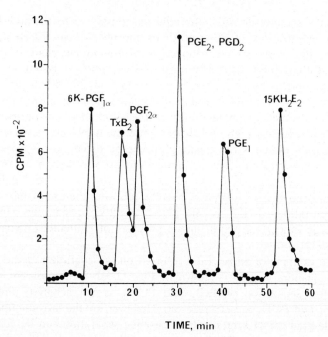

Fig. 11.5.2. HPLC separation of prostaglandins. Chromatographic conditions: column, μBondapak fatty acid reversed phase column; mobile phase, water–acetonitrile–benzene–acetic acid (76.7:23.0:0.2:0.1); flow rate, 2 ml/min; detection, one-minute fractions were collected and assayed for radioactivity. The peaks shown represent 1–2 μg or 0.05–1.0 μCi of tritiated compound. $15KH_2E_2$, 15-keto-13,14-dihydro PGE_2. Reproduced from Whorton et al. (1979), with permission.

was first described in TLC, where silica impregnated with silver nitrate was used to resolve prostaglandins of the 1, 2 and 3 series (Green and Samuelsson, 1964). Argentation HPLC for the separation of prostaglandins was first described in 1976 for the identification of traces of 5-*trans*-$PGF_{2\alpha}$ in samples of PGE_2 (Merritt and Bronson, 1977). In argentation HPLC silica cation-exchange stationary phases which have been treated to form phenylsulphonic acid derivatives are converted from their usual Na^+ form to the required Ag^+ form by washing with 1 M silver nitrate (Powell, 1982).

Retention in argentation chromatography is thought to occur by two separate but simultaneous mechanisms: firstly through interaction of the stationary phase silver ions with the olefinic double bonds of the sample molecules, thereby making retention dependent on the degree of unsaturation of the sample, and secondly through interaction of the polar groups of the stationary phase with the polar groups on the sample molecule.

The most popular mobile phases used in argentation HPLC have been mixtures of chloroform (80–90%) and acetic acid (0.5%) with varying proportions of acetonitrile and/or methanol to optimally adjust chromatographic retention. Addition of methanol to the mobile phase is thought to reduce interactions with polar groups on the stationary phase (Powell, 1982) while acetonitrile reduces retention by direct interaction with silver ions bound to the stationary phase, thereby reducing interactions with the olefinic unsaturated bonds (Merritt and Bronson, 1977). Adjustment of the composition of the mobile phase allows the specific types of interactions between the mobile phase, sample molecules and stationary phase to be enhanced or reduced; in this way an optimal chromatographic separation can be achieved.

An example of the potential of argentation HPLC in the separation of various isomeric forms of a prostaglandin is shown in Fig. 11.5.3 where p-nitrophenacyl esters of 8-*iso*-PGE$_2$, 11-*epi*-PGE$_2$, 5-*trans*-PGE$_2$, PGE$_2$ and PGF$_{1\alpha}$ have been separated. It is to be noted that compounds with *trans* double bonds have shorter retention times than their corresponding *cis* isomer, presumably because the latter have a greater affinity for silver ions. Argentation HPLC may also be used in the separation of isotopic species of prostaglandins; thus both deuterated and tritiated species of a prostaglandin may be resolved (Powell, 1982). The differences in retention between the various isotopic species is thought to reside with the different lengths of the carbon–hydrogen, carbon–deuterium and carbon–tritium bonds. Thus, as the carbon–tritium bond is the shortest of the three species, tritiated samples tend to show stronger interactions with the bound silver ions and therefore longer retention times.

In summary, argentation HPLC provides a useful alternative to conventional reversed phase chromatography for the isolation and quantitation of prostaglandins. The major advantages of this technique are that it provides a better resolution of compounds which

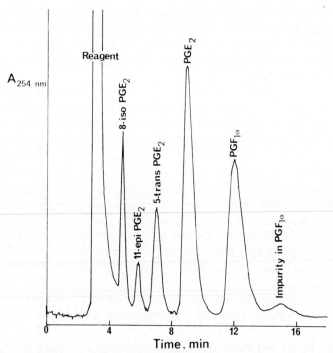

Fig. 11.5.3. HPLC separation of *p*-nitrophenacyl esters of selected prostaglandins. Chromatographic conditions: column, silver ion-loaded Partisil SCX column (250 × 4.6 mm I.D.); mobile phase, 0.06% acetonitrile in dioxane; detection, UV at 254 nm. Reproduced from Merritt and Bronson (1977), with permission.

differ only in their degree of unsaturation or where *cis–trans* isomers of the same compound are present. Furthermore, as the mobile phases use volatile organic solvents, the technique may be conveniently used for the preparation of prostaglandins.

The major disadvantage of the technique is that impurities can accumulate on the column resulting in increasing column back pressure, necessitating not only a column washing procedure, but also regeneration of the column with silver nitrate. A second drawback of argentation HPLC is that with certain compounds, including TxB$_2$ and 13,14-dihydro-15-*oxo*-PGF$_{2\alpha}$, broad or multiple peaks are obtained, together with relatively short retention times (Powell, 1982). In summary, however, the technique can provide a useful alternative to either normal phase or reversed phase chromatography.

11.5.2. Detection of prostaglandins and derivatives

The simplest method of detection of prostaglandins is to measure their UV absorbance at 192 nm which is the wavelength of their maximum absorbance, although PGA_2, 15-keto-PGE_2 and PGB_2 also have prominent maxima at 217, 228 and 278 nm (Hamilton and Karol, 1982), which can be useful for their identification. The minimum detectable quantity by UV absorption is obviously dependent on several factors, including the sensitivity of the detector, the purity of the sample and the UV transparency of the mobile phase, but under optimal conditions the limit of detection is approximately 10 ng. Techniques to improve the sensitivity of UV detection have relied upon derivatisation procedures, which allow measurement at wavelengths higher than 192 nm since the absorbance of mobile phases at the lower wavelengths can be a problem. Several different derivatives can be prepared to allow UV detection at 254 nm:

(a) 2-Naphthacyl esters, which are reported to provide a lower limit of detection of approximately 4 ng (Cooper and Anders, 1974).

(b) p-Nitrophenacyl esters; although no lower limit of detection is quoted, it is probably in the low nanogram range (Merritt and Bronson, 1977).

(c) p-Bromophenacyl esters with a lower limit of detection below 3000 ng (Fitzpatrick, 1976).

(d) p-Nitrobenzyloximes with a lower limit of detection between 25 and 225 ng depending on the number of double bonds present in the prostaglandin (Fitzpatrick et al., 1977).

(e) Finally, the derivatisation procedure providing the highest UV sensitivity is to form an ester with p-(9-anthroyloxy)-phenacyl bromide (trivial name, panacyl bromide). The lower limit of UV detection of the eicosanoid esters is reported to be 280 pg (Watkins and Peterson, 1982).

The panacyl esters described above may also be detected by fluorimetry using an excitation wavelength of 249 nm with a 413 nm cut-off filter (Watkins and Peterson, 1982). Under these conditions the lower limit of detection of the eicosanoid esters is reported to be approximately 50 pg. Two other derivatisation agents which have been reported to allow fluorimetric detection are 7-bromo-7-methoxycoumarin (Turk et al., 1978) and 4-bromomethyl-7-acetoxycoumarin (Tsuchiya et al., 1981).

Refractive index detectors have been used to measure microgram quantities of prostaglandins (Anderson and Leovey, 1974) and recent technological advances should allow a considerable increase in the sensitivity of detection of prostaglandins using this method.

Finally, where either in vivo or in vitro techniques are being applied to monitor the influence of certain variables on prostaglandin biosynthesis, radioactively labelled [^{14}C]- or [^{3}H]-arachidonic acid may be added prior to the initiation of the experiment as a prostaglandin precursor. Using this technique, the labelled arachidonic acid is incorporated into the synthesised prostaglandins and can then be detected either directly using a radioactive flow monitor, or indirectly using a fraction collector with subsequent scintillation counting. The limits of detection utilising radioactive monitoring may be reduced below the picogram level by using high specific activity arachidonic acid as the precursor. Furthermore, the use of radioactive tracers obviates the need for derivatisation procedures prior to chromatographic analysis.

11.5.3. Separation and detection of hydroxyeicosatetraenoic acids (HETEs) by HPLC

Three different chromatographic modes have been applied for the chromatographic separation of HETEs including reversed phase, normal phase and silver ion-loaded cation-exchange. The techniques and sensitivity limits for the detection of the HETEs are generally similar to those reported for the prostaglandins (Section 11.5.2) although UV monitoring at 235 nm is possible because of the presence of a conjugated diene. However, it should be recognised that due to the presence of non-conjugated groups, they also have a second absorbance maximum at 190 nm.

Reversed phase HPLC is undoubtedly the most popular technique for the separation of HETEs and the conditions used for their chromatography are essentially similar to those used for the prostaglandins; however, since the HETEs are less polar, it is necessary to increase the concentration of organic modifier in the mobile phase to facilitate their elution. The precise concentration of organic modifier is dependent upon the specific stationary phase utilised and the precise separation being effected; for example, using an Ultrasphere ODS column, a mobile phase of acetonitrile–0.1% aqueous acetic acid (56 : 44) is reported to elute 15-HETE, 12-HETE and 5-HETE in the sequence given (Peters et al., 1983). Similarly, Eling et al. (1982) used a Radial Pak-A C_{18} cartridge to separate 15-HETE, 12-HETE, 11-HETE and 5-HETE in the elution order shown using a mobile phase of acetonitrile–water (1 : 1), pH 3.5. A mobile phase containing methanol–water–acetic acid (65 : 35 : 0.06), pH 5.3 with a Techsphere 5 C_{18} column eluted 15-HETE, 11-HETE, 12-HETE, 9-HETE and 5-HETE with retention times of 35–50 min and the 5,12-di-HETE isomers eluted between 10 and 15 min (Osborne et al., 1983).

Although infrequently used, good chromatographic separations of the HETEs can be obtained using normal phase chromatography. A separation on a μPorasil silica gel column with a linear mobile phase gradient from 100% hexane–acetic acid (100 : 0.8) to 100% chloroform–acetic acid (100 : 0.8) has been reported to resolve 12-HETE, 11-HETE, 9-HETE and 5-HETE; unfortunately, 15-HETE coeluted with 12-HETE and 8-HETE with 9-HETE (Boeynaems et al., 1980). A second chromatographic separation describes the resolution of 5-HETE, 8-HETE, 9-HETE, 11-HETE and 12-HETE produced by human neutrophils in the presence of arachidonic acid (Goetzl and Sun, 1979); the separation used a normal phase column with a linear mobile phase gradient from 100% hexane–acetic acid (125 : 1) to chloroform–methanol–acetic acid (125 : 5 : 1).

Good resolution of the monoHETEs from the prostaglandins can be achieved on the silver ion-loaded cation-exchange columns which were described earlier (Section 11.5.1). Using these columns with a mobile phase of methanol–acetic acid (99.8 : 0.2), 5-HETE, 11-HETE, 12-HETE and 15-HETE were completely resolved although 8-HETE and 9-HETE were not completely separated (Powell, 1982).

11.5.4. Separation of leukotrienes by HPLC

The vast majority of separations described for the leukotrienes have used isocratic reversed phase conditions with an ODS stationary phase, although a single report exists which used ion-pair reversed phase chromatography with gradient conditions. Examples of these separations will not be discussed, together with some of the problems associated with HPLC of the leukotrienes.

The most popular mobile phases for the separation of the leukotrienes all contain methanol; for example, a mobile phase of methanol–water–acetic acid–ammonium hydroxide (67 : 33 : 0.08 : 0.04), pH 6.2 in combination with an ODS column was used to resolve leukotrienes B_4, C_4, 11-*trans*-C_4 and D_4 in the elution order given (Metz et al., 1982) (Fig. 11.5.4). Under these conditions, the major leukotrienes were well resolved from the diHETEs which are often co-synthesised in in vitro test systems. The addition of acetic acid and ammonium hydroxide to the mobile phase is reported to maximise the sharpness of the eluting peaks and enhance the repro-ducibility of k' and α values (Metz et al., 1982). The mechanisms by which the ammonium and acetate ions improve chromatography was postulated to be through ion-pairing of the free carboxyl and amino functions on leukotrienes C_4 and D_4.

The same authors have reported that overnight flushing of the column with 3% disodium EDTA enhances the recovery of leukotriene C_4 from the stationary phase and that this improvement may be maintained by a daily injection of EDTA through the injector guard column and stationary phase prior to chromatography. The explana-tion of the EDTA effect was proposed to be through the removal of cations which had been retained on the stationary phase and which would otherwise interact with negatively charged moieties on the leukotrienes resulting in either the destruction of their chromophore or the retention of a proportion of the leukotriene molecules on the column which were not ion-paired. Similar column conditioning tech-niques have been suggested where there was a deterioration in the chromatography of the leukotrienes (Anderson et al., 1983). These authors advocated washing of the column with 0.5% disodium EDTA solution in water–methanol (90 : 10) and in some instances recom-

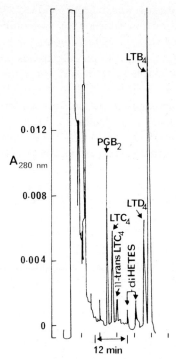

Fig. 11.5.4. HPLC separation of leukotriene standards. Chromatographic conditions: column, 5 μm ODS (250 × 4.6 mm I.D.) coupled to a guard column packed with Corasil (ODS) pellicular packing; mobile phase, methanol–water–acetic acid–ammonium hydroxide (67 : 33 : 0.08 : 0.04), pH 6.2; flow rate, 1 ml/min; detection, UV at 280 nm. The peaks shown include two non-enzymatically formed Δ^6-*trans*-leukotriene B$_4$ (diHETE) isomers of leukotriene B$_4$. Reproduced from Metz et al. (1982), with permission.

mended a second wash with dimethylformamide. It should be empha-sised that removal of either the EDTA or dimethylformamide by flushing the system with HPLC-grade distilled water is essential prior to any chromatographic procedures being carried out.

In addition to providing an adequate separation of the leukotrienes from diHETEs, the use of methanol-based mobile phase systems has also been observed to provide good resolution of leukotriene C$_4$ and D$_4$ from the monoHETEs (Osborne et al., 1983). These authors used a mobile phase or methanol–water–acetic acid (65 : 35 : 0.08) which was adjusted to pH 5.3 by the addition of ammonium hydroxide. The stationary phase used was an ODS column.

Where a separation of both prostaglandins and leukotrienes is required in the same chromatogram and where the detection systems are UV, it is useful to use acetonitrile in the mobile phase due to the relative opacity of methanol at 192 nm; the absorption maximum for the prostaglandins. One such system for the resolution of the major prostaglandins, monoHETEs and leukotrienes synthesised by macrophages has been reported (Peters et al., 1983). The stationary phase was an ODS column and the mobile phases were 0.1% (v/v) aqueous acetic acid, pH 3.7 (solvent A) and acetonitrile (solvent B). Using a gradient system, the major prostaglandins were eluted at 36% acetonitrile and the leukotrienes at 50% acetonitrile, the elution order being leukotriene C_4, E_4, B_4 and D_4. The monoHETEs were eluted at an acetonitrile concentration of 56% and arachidonic acid was eluted at the end of the gradient. It was noted that a rigid control of the mobile phase pH was essential to maintain reproducible retention times for leukotrienes C_4, D_4 and E_4.

An ion-pair separation of the leukotrienes has been reported to facilitate the resolution of leukotriene B_4, (5S, 12S)-6-*trans*-leukotriene B_4, (5S, 12R)-6-*trans*-leukotriene B_4, leukotriene C_4 and D_4 (Ziltener et al., 1983). The initial mobile phase was 70% methanol–30% water–0.1% acetic acid (v/v) with 2.5 mM 1-pentanesulphonic acid as the counter-ion (final pH 3.9). The final mobile phase was 100% methanol–0.1% acetic acid (v/v), and 2.5 mM 1-pentanesulphonic acid, pH 3.9. The stationary phase was an ODS column. The addition of the counter-ion was reported to reduce the long retention times of leukotriene C_4 and leukotriene D_4 at pH 5.7 which was the optimal pH for chromatographic resolution.

The chromatography of leukotriene A_4 presents certain problems since it contains an allylic 5,6-epoxide which breaks down almost instantaneously at neutral or acidic pH. To overcome this chemical instability, alkaline chromatographic conditions are required which use a mobile phase of acetonitrile–0.01 M borate buffer, pH 10 (30 : 70) in combination with either an octyl or ODS column (Wynalda et al., 1982). As silica-based columns are somewhat labile at alkaline pH, careful column maintenance together with relatively frequent column replacement is necessary to maintain chromatographic performance. The use of PRP-1 columns, which are stable at alkaline pH (Section 11.5.1), may provide a solution to these problems of column deterioration.

11.5.5. Detection of leukotrienes

The leukotrienes possess four double bonds including a conjugated triene grouping which is a strong chromophore providing an absorbance maximum between 270 nm and 280 nm, depending on the leukotriene. Thus, leukotriene A has an absorbance maximum at 279 nm (Radmark et al., 1980), leukotriene C at 280 nm and leukotriene D at 270 nm (Murphy et al., 1976). As fixed wavelength detectors with 280 nm filters are relatively common, it is usually at this wavelength that the leukotrienes are monitored.

An alternative strategy for detection, where cellular systems or enzymatic fluxes are being investigated, is the use of radioactively labelled arachidonic acid or glutathione, which can function as precursor molecules and facilitate detection of the synthesised radiolabelled leukotrienes. The labelled molecules can be detected either directly by a radioactive flow monitor, or indirectly by fraction collection and scintillation counting.

Steroids

The steroids are comprised of a very large group of compounds whose chemical properties differ widely. Until the last five years the techniques used for the analysis of steroids were dominated by thin layer chromatography (TLC) and gas–liquid chromatography (GLC) and only comparatively recently has HPLC emerged as an alternative technique. Several authoritative and comprehensive reviews have been written on the chromatographic analysis of the steroids and the reader wishing to gain a detailed knowledge of steroid analysis is referred to these excellent articles (Heftman and Lin, 1982; Kautsky, 1982; Heftman, 1983). In this section the HPLC analysis of the steroids which are generally regarded to be of current clinical importance will be discussed, with subsections relating to sterols and steroid hormones (including oestranes, pregnanes, androstanes and corticosteroids).

11.6.1. Sterols

The sterols include that group of compounds which contain an alcoholic hydroxyl group at C3 and a branched aliphatic chain of at least eight carbon atoms at C17. These compounds, which include cholesterol and its derivatives, can occur as free alcohols or as long-chain fatty acid esters. A number of HPLC techniques have been used in the analysis of the sterols including reversed phase, non-aqueous reversed phase, normal phase, argentation and combinations of the above.

Most of the reversed phase separations have been carried out on ODS stationary phases together with aqueous–organic mobile phases. The major advantage of reversed phase techniques for eliciting sterol separations is that resolution can be achieved both on the basis of carbon number and also on the number of carbon–carbon double bonds. Using a mixture of acetonitrile–water (88:12), separations were achieved which were dependent on the presence or absence of: a Δ^{24}-bond in the side-chain (e.g. Δ^5-cholesterol resolved from $\Delta^{5,24}$-cholestadienol); a C4 methyl group (e.g. 4-methyl-Δ^8-cholesterol resolved from Δ^8-cholesterol); compounds with methyl groups at C4 have also been resolved from those with one methyl group at C4. However, these conditions were unable to resolve compounds on the basis of the presence or absence of a C14 group and there was no resolution of the isomeric C27 sterols with one carbon–carbon double bond although these separations could be achieved using normal phase chromatography, as we shall discuss shortly (Hansbury and Scallen, 1982).

For many of the separations which are found to be unsatisfactory using reversed phase chromatography, resolution can often be obtained using normal phase chromatography on silica stationary phases. For example, using a mobile phase of 2,2,4-trimethylpentane-cyclohexane–toluene (5:3:2) it was possible to resolve isomeric C27 sterol acetate esters with one carbon–carbon double bond; similarly, sterol acetates differing by the presence or absence of a methyl group at C24 were well resolved (Hansbury and Scallen, 1982). However, the same authors observed that unlike reversed phase chromatography, where 4,4-dimethyl sterols were resolved from 4,14-dimethyl sterols, the corresponding sterol acetates were not separated by normal phase chromatography.

Similar chromatographic conditions using two silica columns in tandem together with simple binary solvent systems of hexane–2-propanol (99:1, 24:1, or 9:1), hexane–butanol (99:1) or 2,2,4-trimethylpentane–2-propanol (499:1) have been used for the resolution of a number of the auto-oxidation products of cholesterol (Ansari and Smith, 1979). The authors concluded that while the required separations could be achieved by normal phase chromatography, the use of ODS stationary phases offered added versatility when investigating polar cholesterol auto-oxidation products.

The technique of non-aqueous reversed phase (NARP) chromatography has been described for the quantitation of total cholesterol levels in serum using an ODS stationary phase and a mobile phase of 2-propanol–acetonitrile, although conventional aqueous reversed phase techniques were used for the determination of free and esterified cholesterol in serum (Duncan et al., 1979). NARP chromatography has also been used in the separation of cholesterol and a number of cholesteryl esters (Fig. 11.6.1) with complete resolution being obtained for esters differing in only one methylene group. In general, separation was dependent on both carbon number and number of double bonds; thus esters with higher carbon numbers eluted later while for esters of equivalent carbon number those with the highest

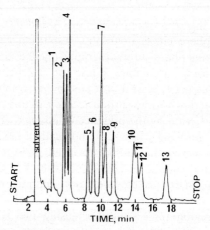

Fig. 11.6.1. HPLC separation of cholesterol and cholesteryl ester standards. Chromatographic conditions: column, Supelcosil LC-18 (250×4.6 mm I.D.); mobile phase, acetonitrile–methanol–chloroform (1:1:1, v/v/v); flow rate, 1.0 ml/min; temperature, ambient; detection, differential refractometer. Peaks: 1, cholesterol, 2, acetate; 3, propionate; 4, butyrate; 5, nonanoate; 6, decanoate; 7, arachidonate; 8, laurate; 9, linoleate; 10, oleate; 11, elaidate; 12, palmitate; 13, stearate. The average mass of lipid chromatographed was 20–40 μg. Reproduced from Perkins et al. (1981), with permission.

number of double bonds eluted first. It was also observed that the *cis* isomers eluted ahead of the corresponding *trans* isomers (Perkins et al., 1981). NARP separations have also been reported for the separation of a number of C27, C28 and C29 sterols including ergocalciferol, sitosterol, stigmasterol, campesterol, cholesterol and ergosterol using a very long ODS column (16 feet × 0.8 in I.D.) with a mobile phase of 0.5% 2-propanol in hexane (Hunter et al., 1978).

Argentation chromatography has been used to obtain an efficient separation of a series of closely related C27 sterol precursors of cholesterol which differed by the number and location of carbon–carbon double bonds (Pascal et al., 1980). In this chromatographic separation the stationary phase was an alumina column which had been loaded with silver nitrate and the mobile phase contained an initial solvent mixture of hexane–toluene (9 : 1) which was changed to (6 : 4) approximately half-way through the separation (Pascal et al., 1980). Argentation chromatography using chloroform to elute a silver nitrate-loaded silica stationary phase has also been used to resolve the *cis* and *trans* isomers of sterol acetates, although column efficiencies were reported to be poor (Colin et al., 1979).

In summary, for the chromatographic resolution of sterols it is advisable that reversed phase techniques should be tried initially and in the event of co-elution normal phase chromatography on silica should be tried as an alternative option. For very specific separations where the aforementioned techniques have failed, either NARP or argentation chromatography could be attempted although it is probable that if reversed phase or normal phase failed to elicit a separation, the remaining techniques would also be unsuccessful.

Sterols are generally monitored with either a refractive index detector or a UV detector. Refractive index detectors are generally rather insensitive and therefore UV detectors have generally been used where limits of detection are critical. The limits of detection at 254 nm for lanosterol, β-stigmasterol and cholesterol have been reported to be approximately 600 ng and for ergosterol 40 ng; however, at 282 nm the limits of detection for the latter are reduced to 700 pg, putting the range of detection in the same order as that reported for GLC with flame ionisation detection (Colin et al., 1979).

11.6.2. Steroid hormones

Several groups are included in this class of steroids including corticosteroids, pregnanes, androstanes and oestranes. Extensive reviews on the chromatography of these compounds have been previously published (Heftman, 1983; Heftman and Lin, 1982; Purdy et al., 1982; O'Hare and Nice, 1982; Schmidt, 1982). It is not the purpose of this section to provide a comprehensive review of the literature but to provide the potential chromatographer with a number of alternative strategies for the separation and detection of the various steroid hormones. For ease of reference the individual classes of steroid hormones have been separated into individual sub-sections.

11.6.2.1. Corticosteroids

The concentration of corticosteroids in the plasma and urine is an indicator of adrenal function and therefore their accurate measurement is of great clinical importance. Throughout the past ten years there has been an increasing trend for clinical laboratories to determine corticosteroid concentrations using HPLC techniques. The most popular chromatographic modes utilised in these determinations include liquid-solid chromatography on silica and reversed phase chromatography.

The first liquid–solid separation of the corticosteroids was carried out using a silica stationary phase with a mobile phase of chloroform–dioxane (100 : 5) to separate cortisol, cortisone and 11-deoxycortisol (Touchstone and Wortmann, 1973). Most separations using silica stationary phases have used variations of the mobile phase originally described by Hesse and Hovermann (1973). Thus, a mobile phase of dichloromethane–ethanol–water (93.6 : 4.7 : 1.7) has been used to resolve prednisolone, cortisol, prednisone, cortisone, corticosterone, deoxycortisol and 17-α-hydroxy-progesterone on a silica stationary phase (Trefz et al., 1975). Other mobile phases which have been used include: hexane–methylene chloride–ethanol–acetic acid (63.8 : 30 : 6 : 0.2) for the determination of plasma prednisolone (Loo et al., 1977), a mixture of 1.5% methanol and 0.2% water in chloroform for the measurement of cortisol in urine (Schwedt et al., 1977), while a mobile phase of dichloromethane–hexane–ethanol–acetic acid (69 : 26 : 3.4 : 1) has been used for the separation of cortisol and methylprednisolone (Ebling et al., 1984). A microbore HPLC column (145 × 0.5 mm I.D.) containing a silica stationary phase, together with a mobile phase of dichloromethane–methanol (97 : 3) has been used for the separation of corticosterone, cortisone and cortisol (Ishii et al., 1978).

Several isocratic systems using reversed phase HPLC with ODS stationary phases have been described; these include mixtures of methanol–water (45 : 55 or 60 : 40) which were used in the determination of the plasma concentrations of cortisol and 11-deoxycortisol (Scott and Dixon, 1979; Reardon et al., 1979) and a mixture of acetonitrile–water (30 : 70) which was used to separate 20α- and 20β-, 5α- and 5β-dihydrocortisol, cortisol and cortisone (Ulick et al., 1977). An isocratic system using a mobile phase of water–acetonitrile–tetrahydrofuran–1 M phosphate buffer, pH 4.4 (63 : 28 : 8 : 1) coupled with a detection system which monitored acid-induced fluorescence has been described for the measurement of plasma cortisol concentrations (Gotelli et al., 1981). When a separation of the corticosteroids and their polar metabolites was required a gradient system using reversed phase HPLC was described. The stationary phase was ODS and, together with an initial mobile phase of methanol–0.01 M ammonium acetate, pH 6.9 (1 : 9) and a final mobile phase of 100% methanol, it was possible to resolve cortisol and a series of cortisol-derived metabolites including glucuronides using the same chromatographic conditions (Slikker et al., 1982). Similarly using an ODS stationary phase with an exponential gradient of methanol–water (40–100% methanol) a resolution of most of the known corticosteroids and their metabolites was achieved (Fig. 11.6.2). Other modes of chromatography which have been used in the separation of the corticosteroids include:

(a) Normal phase chromatography using a nitrile stationary phase with a mobile phase of dichloromethane–2-propanol–water (97.5 : 2.3 : 0.2) which was used in the determination of cortisol in plasma (Van den Berg et al., 1977)

(b) High performance liquid affinity chromatography for the purifi-

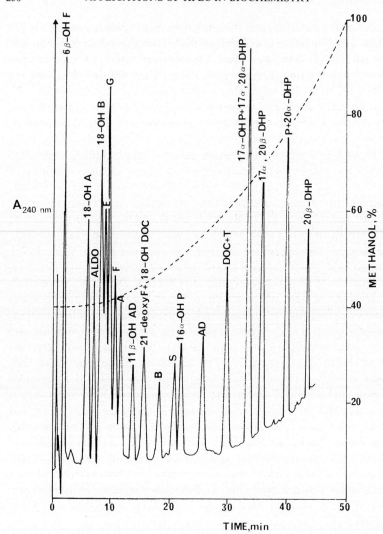

Fig. 11.6.2. HPLC separation of steroid standards. Chromatographic conditions: column, Hypersil ODS (150×4.6 mm O.D.); mobile phase, concave gradient (dashed line) of methanol–water (40–60% methanol); flow rate, 1 ml/min; temperature, 45°C; detection, UV at 240 nm. Peaks: A, 11-dehydrocorticosterone; AD, androstenedione; ALDO, aldosterone, B, corticosterone; DHP, dihydroprogesterone; DOC, 11-de-oxycorticosterone; E, cortisone; F, cortisol; G, adrenosterone, P, progesterone; S, 11-deoxycortisol; T, testosterone. Reproduced from O'Hare and Nice (1982), with permission.

cation and quantitation of cortisol from plasma or urine where the stationary phase consisted of an anti-cortisol antibody which was covalently linked to a silica support. After the cortisol had been bound to the stationary phase it could be reversibly eluted with a high ionic strength buffer (Nilsson, 1983).

Detection of the corticosteroids is facilitated by the strong absorbance of their characteristic 4-ene-3-one resonance structure, which has a maximum around 240 nm (Engel, 1963). At 240 nm the lower limit of detection of the corticosteroids is approximately 2–5 ng; however, using a fixed wavelength detector with a filter with a cut-off wavelength of 254 nm, the limit of detection is increased to approximately 10 ng.

The acid-induced fluorescence of corticosteroids has been used to enhance the detection limit for cortisol in serum. Using a mobile phase buffer at pH 4.4 and incubation of the steroid in ethanolic sulphuric acid prior to chromatography together with excitation and emission wavelengths of 366 nm and 488 nm, respectively, the detection limit is reduced to 500 pg (Gotelli et al., 1981). Several derivatisation procedures are also available to facilitate the fluorescent detection of the corticosteroids and some of these include the use of dansylhydrazine (Kawasaki et al., 1979; Goehl et al., 1979), glycinamide (Seki and Yamaguchi, 1983) and benzamidine (Seki and Yamaguchi, 1984b).

11.6.2.2. Progestins

This group of steroid hormones is characterised by the same Δ^4-3-keto grouping which is found in the corticosteroids and in general the detection methods described for that class of compounds may be applied to the progestins. The analysis of progestins by HPLC has been excellently reviewed (Purdy et al., 1982) and for details of all of the various separation techniques the interested reader is referred to this text. The majority of chromatographic separations used for the resolution of the pregnanes have utilised normal phase and reversed phase chromatography. Normal phase chromatography using an unmodified silica stationary phase and a mobile phase of ethanol–dichloromethane (0.25 : 99.75) was used in the separation of 3 diones (including progesterone) and 6-monohydroxymonoketones (including pregnenolone) of the pregnane series (Lin et al., 1980). Using the same stationary phase, but with a mobile phase of *n*-hexane–methanol–ethanol (96 : 3 : 1) two pairs of pregnane derivatives which were epimeric at C20 were successfully resolved (although the authors noted that reversed phase chromatography gave better resolution). The pairs separated included 20α- and 20β-hydroxy-4-pregnan-3-one and also 5-pregnane-3β, 20α-diol and 5-pregnane-3β-20β-diol (Lin et al., 1980). The same authors reported that resolution of the Δ^5 steroids from their 5α analogues was best performed by reversed phase HPLC using an ODS stationary phase with a mobile phase of acetonitrile–water (60 : 40). Similar separations of 4 epimeric pairs of pregnanediols have been reported using ODS stationary phases with a mobile phase of ethanol–water (55 : 45) (Purdy et al., 1982). A separation of progesterone from its major metabolites was reported by the same authors using a normal phase technique with a Chromegabond diol column (glycophase-type product with *cis* hydroxyl groups on the carbon backbone) and a mobile phase which consisted of a concave gradient of 2-propanol–hexane (initially 0.5 : 99.5, finally 7 : 93).

In summary, an effective resolution of even the most complex mixtures of steroids may be achieved by using a combination of normal and reversed phase chromatography.

11.6.2.3. Androstanes

Both reversed phase and normal phase chromatography have been used for the separation of the androstanes. In general normal phase chromatography using unmodified silica as the stationary phase is the preferred mode of separation since better resolution of all the epimeric steroids except the 17-epimers can be achieved (Hunter et al., 1979). An example of the use of normal phase chromatography to resolve a mixture of androgens is shown in Fig. 11.6.3. Resolution of the epimers of the monohydroxy metabolites of testosterone and androstanedione has been achieved using a mobile phase of hexane–tetrahydrofuran–2-propanol (80 : 15 : 5) with a silica stationary phase (Kawalek et al., 1981).

The ability of reversed phase HPLC to resolve the androstanes is

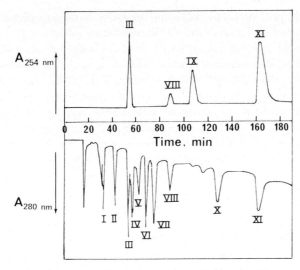

Fig. 11.6.3. HPLC separation of androgens. Chromatographic conditions: column, two Partisil 5 columns in series (300 × 4.6 mm I.D.); mobile phase, dichloromethane–acetonitrile–2-propanol (179 : 20 : 1); flow rate, 0.4 ml/min; temperature, ambient; detection, UV at 254 nm and 280 nm. Peaks: I, androstanedione, II, etiocholanedione; III, androstenedione; IV, epietiocholanolone; V, dehydroepiandrosterone; VI, epiandrosterone; VII, androsterone, VIII, androstadienedione; IX, testosterone, X, etiocholanolone; XI, epitestosterone. Reproduced from Hunter et al. (1979), with permission.

somewhat limited and often the use of more than one isocratic mobile phase or a gradient system is required to obtain a reasonable separation. For example, using an ODS stationary phase to separate the major hydroxylated metabolites of testosterone an initial mobile phase of methanol–water (55 : 45) was used to resolve 2β-, 6β-, 7α- and 16α-hydroxytestosterone but a second mobile phase of methanol–water–acetonitrile (55 : 35 : 10) was required to elute testosterone and androstenedione (Van der Hoeven, 1980). However, it was noted that, although baseline separation of 7α- and 6β-hydroxytestosterone was not achieved, the use of an ODS stationary phase with different selectivity characteristics might provide better resolution. The simultaneous separation of four different testicular Δ^4-3-ketosteroids including testosterone, androstenedione, 17α-hydroxyprogesterone and progesterone has been reported using an ODS stationary phase with a mobile phase of either tetrahydrofuran–methanol–water (16 : 28 : 56) or methanol–acetonitrile–water (9 : 36 : 55) (Darney et al., 1983).

The most convenient method of detection of the androstanes is to utilise their characteristic UV absorbance. The Δ^4-3-ketones (including androstenedione and testosterone) have a strong absorption at 254 nm such that as little as 30 ng of testosterone can be detected. The 17-ketosteroids (including androsterone and etiocholanolone) can be monitored at 280 nm but the sensitivity of detection is five-times less than that of the Δ^4-3-ketones (Hunter et al., 1979).

Methods to enhance the UV absorbance of the androstanes have been described and include the use of 2,4-dinitrophenylhydrazine to form steroid hydrazones (Fitzpatrick et al., 1972) and the use of benzoyl or p-nitrobenzoyl chloride to form the corresponding steroid esters (Fitzpatrick and Siggia, 1973). While the sensitivity of detection of the steroids is increased using these systems, the lack of specificity of the reagents can result in the formation of multiple products (O'Hare and Nice, 1982). More recently, conjugated 17-ketosteroids have been determined in biological fluids using dansylhydrazine as a fluorescent labelling agent (Kawasaki et al., 1982). Derivatisation has also been used as a method of allowing the quantitation of steroids by electrochemical detection. Thus, p-nitrophenylhydrazine has been used for the derivatisation and quantitation of 17-ketosteroids in human serum after they had been initially deconjugated (Shimada et al., 1980). This method has been adapted to allow the pre-column derivatisation of 17-ketosteroid sulphates, without the necessity for deconjugation, prior to reversed phase HPLC (Shimada et al., 1984).

11.6.2.4. Oestranes

The accurate quantitation of oestranes is of great importance in clinical chemistry to provide information about the progress of a pregnancy. In the last months of pregnancy the urinary oestriol levels normally rise, and a decline in levels is indicative of a malfunction. The oestranes are usually excreted in urine as conjugates and in most analyses these are hydrolysed prior to HPLC analysis. There are two methods currently in use for hydrolysis, these being acid hydrolysis and enzymatic hydrolysis. In the former technique the sample is heated to 110 °C for 30 min prior to extraction. In enzymatic hydrolysis a buffered solution of enzyme (often *E. coli* β-glucuronidase) is incubated with the sample at 60 °C for 30 min (e.g. Gotelli et al., 1977). The most popular HPLC techniques used for the resolution of the oestranes have been reversed phase and reversed phase ion-pair HPLC and to a lesser extent normal phase chromatography and anion-exchange chromatography. Additionally, reversed phase ion-pair partition chromatography and argentation chromatography have been used. Each of these techniques will now be discussed and this will be followed by a short note on the methods used for the detection of the oestranes.

Where reversed phase or reversed phase ion-pair techniques have been used for the resolution of the oestranes, the stationary phase utilised has usually been ODS together with a mobile phase consisting of an aqueous solution of methanol or acetonitrile. For example, using an ODS stationary phase with a mobile phase of methanol–0.1% aqueous ammonium carbonate (55 : 45), oestriol and oestradiol were successively eluted (Dolphin and Pergande, 1977). The same authors noted that when a normal phase separation was performed using a silica stationary phase and a mobile phase of ethanol–*n*-heptane (5 : 95) the elution order was altered to oestrone, oestradiol and oestriol. In the reversed phase mode the importance of the organic modifier in determining the elution order was demonstrated using an ODS stationary phase and a mobile phase of acetonitrile–0.5% aqueous ammonium dihydrogen phosphate, pH 3.0 (1 : 2) to resolve several 17-keto and 17-hydroxyl oestranes. It was noted that when methanol was substituted for acetonitrile the elution order was reversed (Shimada et al., 1979).

A comprehensive set of HPLC separations for the resolution of 25 conjugated and non-conjugated oestranes has been described (Slikker et al., 1981), which includes the initial separation of a complex mixture using a gradient reversed phase system consisting of an ODS stationary phase with an initial mobile phase of methanol–0.01 M ammonium acetate buffer, pH 6.9 (10 : 90) and a final mobile phase of methanol (Fig. 11.6.4). The separated oestranes were divided into six groups according to their retention time and subsequently subjected

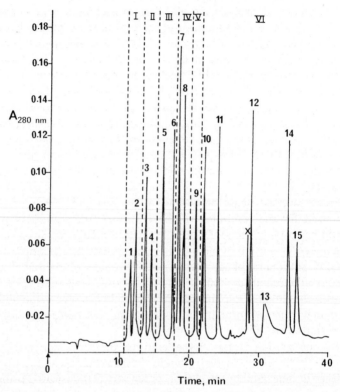

Fig. 11.6.4. HPLC separation of oestrogen and metabolite standards. Chromatographic conditions: column, LiChrosorb RP-18 (5 μm) (250×9 mm I.D.); mobile phase, convex gradient of methanol–0.01 M ammonium acetate, pH 6.9 (10–100% methanol over 50 min); flow rate, 2.0 ml/min at 2,500 p.s.i.; temperature, ambient; detection, UV at 280 nm. Peaks: 1, oestriol-3-β-D-glucuronide (20 μg); 2, oestradiol-3-sulphate, 17-β-D-glucuronide (200 μg); 3, oestradiol-3,17-disulphate (80 μg); 4, oestriol-3-sulphate (100 μg); 5, oestriol-16α-glucuronide (20 μg); 6, oestriol-17-sulphate (10 μg); 7, oestrone-3-glucuronide (40 μg); 8, oestradiol-17-β-D-glucuronide (20 μg); 9, oestrone-3-sulphate (100 μg); 10, 6α-hydroxyoestradiol (10 μg); 11, oestriol (10 μg); ×, unknown degradation product; 12, 17α-oestradiol-17-β-D-glycoside (20 μg); 13, 2-hydroxyoestradiol (15 μg); 14, 17β-oestradiol (10 μg), oestrone (10 μg); 15, 2-methoxyestradiol (5 μg). I–VI, see text. Reproduced from Slikker et al. (1981), with permission.

to further HPLC analysis using four different chromatographic systems to increase their resolution. Two of these systems used a pair of ODS columns linked in series and were isocratic:

(*a*) Using a mobile phase of methanol–0.01 M ammonium acetate, pH 3.97 (45 : 55) the oestranes from the following groups were further resolved: Group I (oestriol 3-glucuronide, oestriol 3-sulphate, 17β-D-glucuronide), Group III (oestriol 16α-glucuronide, oestriol 17β-glucuronide, oestriol 17-sulphate) and Group IV (oestrane 3-glucuronide, oestradiol 3-glucuronide, oestradiol 17β-glucuronide).

(*b*) A mobile phase of methanol–0.01 M ammonium acetate, pH 7.74 (35 : 65) was used to further resolve Group II oestranes (oestradiol 3,17-disulphate, oestriol 3-sulphate).

(*c*) A C_2 column was used with a mobile phase of methanol–0.01 M ammonium acetate, pH 7.56 (25 : 75) for the separation of Group V oestranes (oestrone 3-sulphate, oestradiol 3-sulphate, 15α-hydroxy-oestriol).

(*d*) A normal phase gradient system using a Chromegabond Diol column and a mobile phase which was initially hexane and finally hexane–2-propanol (80 : 20) was used to further resolve the Group VI oestranes (oestrone, oestradiol, oestriol, 2-methoxyoestradiol, 2-hydroxyoestradiol, 6α-hydroxyoestradiol and 17α-oestradiol-17β-D-glycoside).

When normal phase HPLC techniques have been used for the separation of the oestranes the stationary phase has usually been unmodified silica, although Diol columns have also been used (as described earlier). Using a mobile phase of hexane–ethanol (9 : 1) and a silica stationary phase, good resolution of the more polar oestrogens including oestradiol, 6-ketooestradiol, 16-epioestriol, oestriol, 16,17-oestriol, 2-hydroxyoestriol and 6-hydroxyoestriol has been described (Lin and Heftmann, 1981). Similarly, these authors noted that for separations of the less polar oestranes a mobile phase of hexane–ethanol (97 : 3) may be used.

In general, it is suggested that epimers are better resolved using normal phase chromatography, but when separations of compounds which differ by the addition of ketone, hydroxy or carbon–carbon double bond are required, reversed phase chromatography is the technique of choice (Lin and Heftmann, 1981).

Ion-exchange chromatography has been used for the resolution of the oestranes and while in general this technique is much less popular than reversed phase or normal phase chromatography it can nevertheless provide good separations. The stationary phases used have been either silica-based (Musey et al., 1978) or cellulose-based anion-exchange columns (Van der Wal and Huber, 1974, 1977) and the mobile phases have been either 0.01 M potassium dihydrogen phosphate, pH 4.2 or 0.1 M sodium chloride, pH 5.0 (Musey et al., 1978) or perchlorate (varying concentrations) plus 0.01 M phosphate buffer, pH 6.8 (Van der Wal and Huber, 1977). The latter authors noted that in general the elution order was oestriols, oestranes, equilin, oestradiol, 17α-dihydroequilin, equilenin and dihydroequilenin. Also ring A conjugates were eluted before those of ring D, and glucuronide conjugates were eluted before phosphates and before sulphates.

Two further separation techniques have been described for the resolution of the oestranes: reversed phase ion-pair partition (Hermansson, 1978) and argentation reversed phase HPLC (Tscherne and Capitano, 1977). In the former technique C_2 or C_8 columns were coated with 1-pentanol and used with tetrapropylammonium hydroxide as an ion-pairing reagent. In the latter technique an ODS stationary phase was used with a mobile phase of methanol–5% silver nitrate (60 : 40) to elute oestriol, equilin, oestrone and oestradiol.

Several methods of detection have been applied to the determination of the oestranes, of which the most popular are UV, fluorescence (either native or derivatised) and electrochemical. The application of UV to the detection of oestranes has benefited over recent years from the development of more sensitive and stable UV monitors which can detect the characteristic absorbance of the aromatic A ring of the oestranes at 280 nm. The lower limits of detection at this wavelength allow 10 ng of oestriol to be detected. Increased sensitivity of detection is possible at 200 nm but the increased background signal often precludes the use of this wavelength. The formation of derivatives which absorb strongly in the UV is one strategy which has been used to enhance the sensitivity of detection of the oestranes. A variety of derivatisation procedures have been reported including 2,4-dinitrophenylhydrazone (DNPH), which forms derivatives with compounds having a carbonyl group to produce molecules with a strong UV absorbance at 254 nm and 365 nm, and p-iodobenzenesulphonyl esters, which have a strong UV absorbance at 254 nm (Roos, 1976).

Monitoring the natural fluorescence of the oestranes provides a considerable increase in sensitivity over that achieved with UV detection. Using an excitation wavelength of 220 nm, oestranes give a primary emission wavelength of 310 nm and a secondary emission at 608 nm; using the latter, a lower limit of detection for oestriol of 100 pg has been reported (Taylor et al., 1980).

The fluorescence detection of the oestranes can be increased by the formation of derivatives containing highly fluorescent groups. The most popular derivatisation procedure has been dansylation using dansyl chloride (Fishman, 1975) which, with an excitation wavelength of 350 nm and an emission wavelength of 540 nm, increases the sensitivity of detection of oestradiols to less than 50 pg (Roos, 1978) and less than 400 pg for oestriol (Schmidt, 1982).

Electrochemical detection has also been applied to the oestranes with electrode potentials set in the range 0.7–0.86 V versus a Ag/AgCl reference electrode (Shimada et al., 1979, 1981; Prescott et al., 1982). A variety of oestranes can be detected electrochemically including oestriol, oestradiol, oestrone, 2-hydroxyoestrone, 4-hydroxyoestrone, 2-methoxyoestrone, 2-methoxyoestradiol (Shimada et al., 1979) and also 2-hydroxyoestradiol and 4-hydroxyoestradiol (Shimada et al., 1981; Prescott et al., 1982). The limits of detection of the oestranes using electrochemical detection have been reported to be 1 ng of 2-hydroxyoestradiol and 5 ng for oestriol (Shimada et al., 1979).

Biogenic amines

The biogenic amines include a very large number of pharmacologically active compounds and their metabolites and so, for the sake of brevity, this section will only deal with those compounds which are presently implicated in the diagnosis and understanding of a number of pathological disorders, including hypertension, adrenal and neuronal tumors and certain psychiatric disorders. Throughout the last decade there has been a large expansion in the use of HPLC for the determination of the concentrations of the biogenic amines and now a wide variety of HPLC techniques are available for the separation and quantitation of these compounds. Some of the chromatographic modes utilised include cation-exchange, reversed phase, reversed phase ion-pair and reversed phase ion-pair partition chromatography. This array of chromatographic methods has been coupled to an equally wide variety of detection techniques including UV detection, fluorimetric detection using either native fluorescence or coupled with pre- or post-column derivatisation, electrochemical detection and radiochemical detection.

Certain aspects of both the chromatography and detection will be discussed in the following section together with some of the advantages and disadvantages peculiar to each approach.

11.7.1. Chromatography

In this section we shall initially discuss reversed phase chromatography followed by the closely related ion-pair chromatography and ion-pair partition chromatography. Finally, the use of cation-exchange stationary phases in the separation of the biogenic amines will be discussed.

The most popular mode of separation used at the present time is unquestionably reversed phase. The behaviour of biogenic amines on ODS columns has been reported in some detail (Molnar and Horvath, 1976) where it was proposed that retention was determined by the solvophobicity of the compounds (Chapter 7). However, the quality of the chromatography reported in this paper has been suggested to be due in part to ion-pairing between the catecholamines and the phosphate buffer present in the mobile phase (Asmus and Freed, 1979). These authors showed that either nitric acid or sulphuric acid in the mobile phase gave good separations and peak symmetry and also provided retention of very polar biogenic amines; furthermore, the use of trichloroacetic acid gave retention times and separation efficiencies equivalent to that obtained using sodium octylsulphate in an ion-pair chromatographic mode. If separations of this type are attempted, care should be taken in the selection of the stationary phase as poor peak shapes have been reported using octyl-bonded silica where strong interactions between the very polar groups of the catecholamines and non-derivatised sites of the octyl-modified support may occur (Crombeen et al., 1978). Despite these problems, good separations of the o-phthalaldehyde (OPA) derivatives of both the catecholamines and serotonin have been reported with a two-step isocratic system using 0.025 M sodium phosphate buffer, pH 5.1, and an initial organic modifier of 25% (v/v) acetonitrile which was changed after 15 min to 45% (v/v) methanol (Davis et al., 1982). The stationary phase used was a phenyl column and under these conditions the elution order was histamine, noradrenaline, octopamine, dopamine, serotonin and tyramine. Similar reversed phase separations have been described for the trihydroxyindole derivatives of the catecholamines (Kringe et al., 1983) and also underivatised catecholamines and serotonin using a mobile phase of 0.01 M perchloric acid–acetonitrile in ratios of 99 : 1 and 85 : 15, respectively (Jackman et al., 1980). Separation and quantitation of nine different metabolic derivatives of the monoamines has also been described using a mobile phase of 0.07 M sodium phosphate plus 0.1 mM EDTA–methanol (9 : 1), pH 5.4 with an ODS stationary phase (Van Bockstaele et al., 1983). A direct comparison has been made between the efficiency of a reversed phase chromatography column (Zorbax ODS) and a strongly acidic cation-exchange column (Zipax SCX) (Okamoto et al., 1978). The authors concluded that while the ODS column offered a higher number of theoretical plates, it was often difficult to separate certain of the catecholamine peaks from the solvent front and, in addition, that certain peaks could not be resolved from extraneous components in certain instances; for these reasons the authors favoured the cation exchange column.

The difference between reversed phase chromatography and the use of organic acids in an ion-pair reversed phase mode is somewhat poorly defined and so in considering the use of ion-pair chromatography in the separation of the biogenic amines the discussion will be confined to 'soap chromatography'. This technique was originally introduced by Knox and co-workers, who showed that when a mobile phase of methanol–water contained a low concentration of sodium dodecyl sulphonate a separation of the catecholamines and their metabolites was achieved with good retention, efficiency and peak symmetry (Knox and Pryde, 1975; Knox and Jurand, 1976). The stationary phase in these studies was ODS which had been further treated with trimethylsilane to react with any free silanol groups which might otherwise have reduced the separation efficiency.

Currently, the most popular stationary phase used in ion-pair chromatography of biogenic amines is ODS and specially prepared ion-pair columns are now commercially available in which the number of free silanols has been reduced. The ODS stationary phases are generally used with a mobile phase of either citrate or phosphate buffer (pH 2–5), often with an organic modifier (methanol or acetonitrile) and a counter-ion at concentrations between 0.1 and 20 mM. The counter-ions commonly used include sodium dodecyl sulphonate, sodium octyl sulphonate (Honma, 1982), sodium lauryl sulphate (Taylor et al., 1983) and sodium heptane sulphonate (Moyer and Jiang, 1978). An example of the resolving power of reversed phase ion-pair chromatography for the catecholamines and their metabolites is shown in Fig. 11.7.1 where a good separation of 15 different compounds has been demonstrated. The authors noted, however, that the presence of sodium octyl sulphate in the mobile phase decreased the lifetime of the columns.

A chromatographic technique closely related to ion-pair chromatography is ion-pair reversed phase partition chromatography, as originally described by Schill (1974). This technique has been used for the separation of organic ammonium compounds including noradrenaline, adrenaline and dopamine (Johannson et al., 1978). In the separations described, a reversed phase column (either octyl or ODS) was coated with an organic solvent (either methylene chloride or 1-pentanol) to form a liquid stationary phase which was used with a phosphate buffer as a mobile phase. A series of compounds were used as counter-ions including dihydrogen phosphate, cyclohexyl-sulphamate, dicyclohexylsulphamate and octylsulphate. It was noted that a stationary phase of methylene chloride was required to separate the highly hydrophilic noradrenaline and adrenaline, while a long-chain alkylammonium compound was necessary as a counter-ion to prevent peak asymmetry in the chromatography of hydrophilic amines and quaternary ammonium compounds. It was also noted that the elution order on the octyl-bonded support was in accordance with the normal behaviour for ion-pair distribution between an organic phase and an aqueous phase whereas on the ODS stationary phase the column retained certain hydrocarbonaceous characteristics resulting in a reduced separation efficiency compared with the octyl support (Johansson et al., 1978).

Cation-exchange HPLC has been extensively used for the separation of the catecholamines. The elution order of the compounds is dependent on the charge which they carry at the ionic strength and pH of the buffer. While this technique would appear optimal for this series of chemical entities, problems arise in the separations due to the relatively low efficiencies of the cation-exchange columns available. Furthermore, until recently the quality of the commercially available columns was rather variable. However, a number of separations of the catecholamines have been reported using cation-exchange chromatography (Refshauge et al., 1974; Fenn et al., 1978; Keller et al., 1976; Allenmark and Hedman, 1979). More recently the separation of both the glycylglycine derivatives and the 2-cyanoacetamide derivatives of the catecholamines has been demonstrated using cation-exchange HPLC (Seki and Yamaguchi, 1984a; Honda et al., 1983). An example of the potential of cation-exchange HPLC in the separation of the

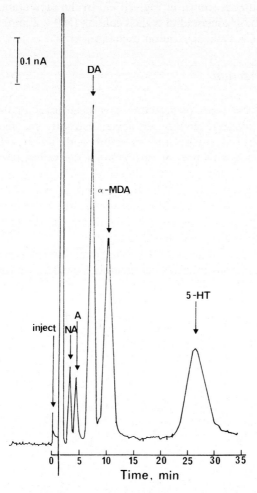

Fig. 11.7.2. HPLC separation of catecholamines and serotonin. Chromatographic conditions: column, Vydac-CX (35 μm) (500×2 mm I.D.); mobile phase, 27.4 mM citrate, 50 mM sodium acetate, 18.4 mM acetic acid, 60 mM sodium hydroxide, adjusted to pH 5.3; flow rate, 0.6 ml/min; temperature, ambient; detection, electrochemical, carbon paste working electrode, electrode potential +0.55 V vs. an Ag–AgCl reference electrode. Peaks: NA, noradrenaline (80 pg); A, adrenaline (100 pg); DA, dopamine (400 pg); α-MDA, α-methyldopamine (internal standard); 5-HT, serotonin (200–250 pg). Reproduced from Patthy and Gyenge (1984), with permission.

biogenic amines is shown in Fig. 11.7.2. In the separation shown the stationary phase consisted of a glass column (500 × 2 mm) dry packed with Vydac-CX pellicular cation-exchanger.

11.7.2. Detection

Several different types of detectors have been used for the quantitation of the biogenic amines; however, currently the most popular techniques are electrochemical detection and fluorescent detection, usually coupled with pre- or post-column derivatisation techniques. The specific details of these detection techniques are discussed elsewhere (Chapter 3) and in the present section only the basic principles as applied to the biogenic amines will be reviewed together with a comparison between the two techniques in terms of their respective advantages and disadvantages.

Electrochemical detection of the biogenic amines relies upon the loss of two electrons from the compound to the detector electrode and

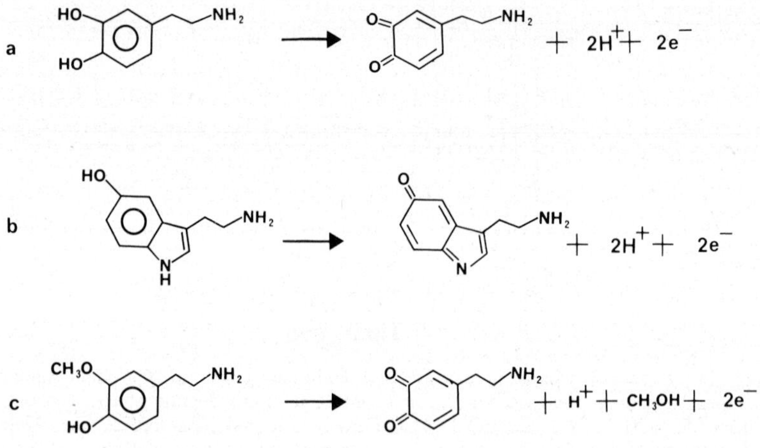

Fig. 11.7.3. a: Typical 2 electron oxidative mechanisms for catecholamines. b: 2 electron oxidation mechanisms for 5-hydroxyindoles. c: 2 electron oxidation mechanisms for O-methylated catechol metabolites. Reproduced from Mefford (1981), with permission.

thereby provides the basis for the detection mechanism (see Fig. 11.7.3). Thus 5-hydroxyindoles such as serotonin are converted to the corresponding imine-quinone while catecholamines are oxidized to quinones. The current derived from electron abstraction is proportional to the concentration of compound in solution. Two types of electrochemical detector have been used in the measurement of the biogenic amines: amperometric detectors, which are the most common, are basically non-destructive, oxidizing approximately 1–5% of the compound while the alternative coulometric detectors attempt to achieve 100% oxidation.

Since the introduction of electrochemical detection in combination with HPLC for the detection of catecholamines in rat brain tissue (Kissinger et al., 1973), the chromatographic literature has been deluged with a host of papers describing modifications either to the chromatographic methods or the extraction procedures. Some of the more important papers in this literature include Keller et al. (1976), Felice et al. (1978), Moyer and Jiang (1978) and Taylor et al. (1983). Electrochemical detection has also been used in the determination of metabolites of the catecholamines (Wightman et al., 1977; Shoup and Kissinger, 1977; Hefti, 1979; Taylor et al., 1983).

The specific electrochemical conditions used for the amperometric detection of the biogenic amines varies according to the specific compound being detected. Thus, the catecholamines and 3,4-dihydroxyphenylacetic acid are usually measured with an electrode potential of $+0.60$ V versus an Ag–AgCl reference electrode, while for serotonin, 5-hydroxyindoleacetic acid, 3,4-dihydroxyphenylacetic acid, 3-methoxy-4-hydroxyphenylglycol and homovanillic acid an applied potential of $+0.85$ V is used (Mefford, 1981). Detection of several other metabolites of the biogenic amines, including 4-hydroxy-3-methylphenylethyleneglycol, 4-hydroxy-3-methoxyphenylacetic acid and 5-hydroxyindole-3-acetic acid, are also detected using an electrode potential of $+0.80$ V versus a Ag–AgCl reference electrode (Taylor et al., 1983).

The minimum limit of detection for the biogenic amines using conventional amperometric electrochemical detectors is generally considered to be in the range of 10–100 pg (Patthy and Gyenge, 1984). A relatively new development in electrochemical detection is the use of dual electrochemical detectors which use two working electrodes (anode and cathode) operated at different potentials (Blank, 1976; Roston and Kissinger, 1981). More recently, a dual electrochemical detector designed for use in microbore HPLC together with a pre-column for enriching catecholamines has been reported to reduce the minimum limits of detection for noradrenaline and adrenaline to approximately 3 pg (Goto et al., 1983). It should, however, be noted that the very sensitive limits of detection described earlier can only be achieved under optimal detector conditions. It has been observed that the ionic strength of the mobile phase is an important parameter in detector response and is optimal at a salt concentration of 0.07 M in an aqueous medium (Moyer and Jiang, 1978). The concentration of methanol in the mobile phase can also affect the sensitivity of detection and it has been reported that at salt concentrations below 0.04 M the presence of methanol dramatically reduces the detector sensitivity (Patthy and Gyenge, 1984).

Fluorimetric detection of catecholamines is subdivided into three major classes, these being native fluorimetry, pre-column derivatisation and post-column derivatisation. Native fluorescence utilises the fluorescence characteristics intrinsic to the native biogenic amine molecules. The catecholamines have excitation maxima at 200–220 nm and 280–300 nm with an emission maximum at 310–330 nm while the indoles have excitation maxima at 210–220 nm and 270–280 nm and an emission maximum at approximately 360 nm (Anderson and Young, 1981). The minimum limits of detection for the catecholamines using native fluorescence are generally considered to be in the range 100–300 pg and rather lower for serotonin (20 pg) (Jackman et al., 1980). Other workers, however, have claimed detection limits of around 20 pg for a series of dopamine metabolites (Schusler-Van Hees and Beijersbergen-Van Henegouwen, 1980).

In pre-column derivatisation the biogenic amine is initially conjugated with a derivatisation agent to produce an adduct which possesses enhanced fluorescence. The adduct is then extracted and chromatographed. The most commonly used derivatisation reagent is o-phthalaldehyde (OPA) although dansyl chloride (Schwedt and Bussemas, 1976, 1977) and fluorescamine (Imai, 1975; Imai and Zamura, 1978) have also been used. A major advantage of OPA derivatives is their known stability, which has been shown to be in excess of 100 h when stored in ethyl acetate at 2–8°C (Davis et al., 1978). Good chromatographic separations of the OPA derivatives of the biogenic amines present in blood, brain tissue and urine have been reported (Jones et al., 1981; Davis et al., 1982). The minimum limits of detection for the biogenic amines from plasma and tissue using OPA derivatisation is considered to be in the range of 50–100 pg (Davis et al., 1982). More recently, a continuous wave argon laser has been used as an excitation source to provide enhanced sensitivity of detection and limits of 5 pg and 16 pg have been claimed for noradrenaline and dopamine, respectively (Todoriki et al., 1983).

Post-column derivatisation techniques for the detection of the biogenic amines rely upon the initial chromatographic separation of the biogenic amines with subsequent on-line derivatisation to form a fluorescent adduct prior to fluorimetric detection. The most commonly used derivatisation methods include the trihydroxyindole (THI) method and the o-phthalaldehyde (OPA) method, although more recently ethylenediamine (Seki, 1978) glycylglycine (Seki and Yama-guchi, 1984a) and 2-cyanoacetamide (Honda et al., 1983) have also been used. The THI method for the HPLC analysis of catecholamines was originally developed by Mori (1974) and with recent methodological improvements a minimum detection limit of 20–30 pg noradrenaline in tissues and urine has been reported (Okamoto et al., 1978). The THI method has been reported to be more selective than the OPA method for the detection of catecholamines in tissue samples, suggesting that there are fewer reactions with co-chromatographing endogenous material, although it should be noted that the OPA method is favoured for the detection of dopamine (Okamoto et al., 1978).

Although not generally used for applications where sensitivity of detection is a limiting factor, UV detection at a wavelength of 280 nm may be utilised. The minimum detection limits for the determination of noradrenaline, DOPA and dopamine have been quoted to be between 5 and 10 ng, i.e. approximately half that achievable using native fluorescence systems (Scratchley et al., 1979).

Of the detection techniques reviewed, electrochemical and native fluorescence are broadly comparable on the basis of cost, ease of maintenance and sample preparation time. Fluorescence methodologies requiring derivatisation are rather more complex with specific problems associated with individual systems. For example, with pre-column derivatisation an initial chemical reaction has to be carried out and the derivative subsequently extracted from the reaction mix for chromatographic analysis. In post-column derivatisation, the chromatographic system needs to be carefully arranged to minimise dead-space between the column and the detector; adequate time also needs to be allowed for the derivatisation reaction to reach completion.

In terms of sensitivity of detection, electrochemical detectors are generally suggested to be approximately four times more sensitive than systems using native fluorescence for the detection of noradrenaline, DOPA and dopamine (Scratchley et al., 1979). However, for certain compounds native fluorescence provides equal or greater sensitivity than electrochemical detection. For example, while electrochemical detection is more sensitive than fluorescence for the measurement of homovanillic acid, it is equally sensitive for 5-hydroxyindoleacetic acid and is less sensitive for the measurement of tryptophan (Anderson and Young, 1981).

It should be recognised that for the measurement of certain compounds both the chromatographic system and the detector utilised need to be carefully considered. For example, electrochemical detectors often show a large solvent front after sample injection. This problem is caused by the sensitivity of electrochemical detectors to changes in the composition of the solvent system (Anderson and Young, 1981) and practically it means that compounds with short retention times may be buried in the solvent front.

In conclusion, the specific detector chosen for the quantitation of the catecholamines is very dependent on the specific application for which the system is to be used. For systems requiring high sensitivity either electrochemical or fluorescence derivatisation systems are probably optimal. If sensitivity is not a problem, then clearly the choice of detector is wider and the selected system could utilise electrochemical, native fluorescence or even UV detection.

Vitamins

11.8.1. Introduction

Vitamins are conveniently classified into two groups which are distinguished by being either water soluble or fat soluble (Table 11.8.1). Conventional purification methods required the initial extraction of large amounts of tissue because of the very low vitamin content and therefore an alternative approach was required which fulfilled the following criteria:

(1) Capable of handling trace quantities without causing significant loss.

(2) Rapid.

(3) Able to avoid heat, UV light and sources of oxidation.

(4) Conveniently coupled to a highly sensitive detection system since many vitamins do not contain useful chromophores.

HPLC fulfills all of these criteria and is now used extensively in the analysis of samples from a variety of sources including mammalian tissue, plant tissue and food extracts. The overall efficiency of HPLC and the variety of chromatographic modes allows the majority of analyses to be performed by either reversed phase, reversed phase ion-pair, normal phase or ion-exchange HPLC. Other chromatographic modes such as size exclusion and affinity have found limited application in the chromatography of the vitamins.

Table 11.8.1

Vitamins and their coenzyme forms

Type	Coenzyme or active form	Function promoted
Fat-soluble		
Vitamin A	11-*cis*-Retinal	Visual cycle
Vitamin D	1,25-Dihydroxychole-calciferol	Calcium and phosphate metabolism
Vitamin E	–	Antioxidant
Vitamin K	–	Prothrombin biosynthesis
Water-soluble		
Thiamin	Thiamin pyrophosphate (TPP)	Aldehyde group transfer
Riboflavin	Flavin mononucleotide (FMN), flavin adenine dinucleotide (FAD)	Electron transfer
Nicotinic acid	Nicotinamide adenine dinucleotide (NAD, NADP)	Electron transfer
Pantothenic acid	Coenzyme A (CoA)	Acyl group transfer
Pyridoxine	Pyridoxal phosphate	Amino group transfer
Biotin	Biocytin	Carboxyl group transfer
Folic acid	Tetrahydrofolic acid	One-carbon group transfer
Vitamin B_{12}	Coenzyme B_{12}	1,2 shift of hydrogen atoms
Lipoic acid	Lipoyllysine	Hydrogen atom and acyl group transfer
Ascorbic acid	–	Cofactor in hydroxylation

Many vitamins are extremely labile and considerable attention has been devoted to improving their stability during extraction and HPLC analysis. As a result of their instability, the major difficulty in the analysis of vitamins by HPLC is to resolve them from the complex molecules found in their environment without causing their degradation. No attempt will be made to cover the entire area of extraction and sample preparation and the reader is referred to a review (Parrish, 1980); however, specific examples are cited in the text where appropriate. References are presented along with specific applications in Section 11.8.4.

11.8.2. Fat-soluble vitamins

11.8.2.1. A vitamins

This class of vitamins comprises a group of molecules which are also known as retinoids and also includes certain carotenes which display vitamin A-like activity (particularly α-, β- and γ-carotene from plants). Many derivatives of these compounds are found naturally and include the alcohol, the acetate and the palmitate. The best chromatographic mode for the resolution of the vitamin A class is often determined by the conditions used for the extraction. For example, saponification to remove lipids is often followed by extraction into solvents, such as hexane, which are compatible with aqueous reversed phase HPLC using either a C_8 or C_{18} silica-based stationary phase (Fig. 11.8.1). An alternative system has been reported using a C_{18} support and a mobile phase of 0.5% acetic acid in acetonitrile (Annesley et al., 1984).

Alternatively, where samples are extracted into organic solvents, normal phase HPLC can also be used. This option is particularly valuable for the resolution of optical isomers of the A vitamins and, for example, 13-*cis*-retinol can be resolved using this system (Egberg et al., 1977). Retinoyl species may be converted to retinol prior to chromatography but the subsequent profile is difficult to interpret because of the number of possible intermediates that may also be generated. A computer-assisted analysis of vitamin A extracts from two species of tobacco leaf using non-aqueous reversed phase is shown in Fig. 11.8.2.

The number of recorded carotene derivatives is increasing as both chromatographic systems and detection methods improve and at present more than 300 naturally occurring carotenoids have been discovered. A recent comparison of the methods available for the assay of the carotenoids concluded that HPLC was superior to other techniques for reasons of speed and ease of quantitation (Krinsky and Wecankiwar, 1984).

Detection of vitamin A and its derivatives is most conveniently performed by monitoring fluorescence; however, UV detection is also

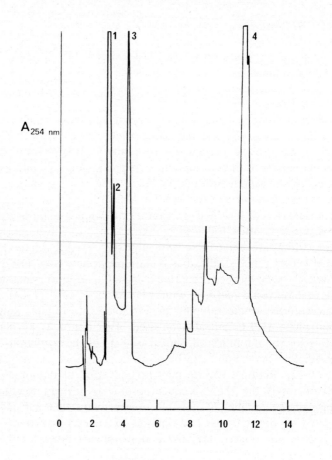

Fig. 11.8.1. Reversed phase separation of vitamin A alcohol and esters. Chromatographic conditions: stationary phase, Zorbax C-8 (250×4.6 mm I.D.); mobile phase, 99.8% water–0.2% perchloric acid (60%), and 99.8% methanol–0.2% perchloric acid; flow rate, 1.6 ml/min; detection, UV at 254 nm. Peaks: 1, vitamin A alcohol; 2, impurity; 3, vitamin A acetate; 4, vitamin A palmitate. Reproduced from Dupont Analytical Systems with permission.

possible at 313–328 nm. Detection of carotenes can be performed using UV detection at 450 nm.

11.8.2.2. D vitamins

HPLC analysis of the D vitamins is particularly difficult because they
are only present at very low concentrations. Many methods of extraction have been used, but the most common is to saponify the sample
and extract it into an organic phase (usually hexane or dimethylsulphoxide). A mobile phase of diethyl ether has been suggested to be
preferable to hexane since this solvent allows an excellent separation

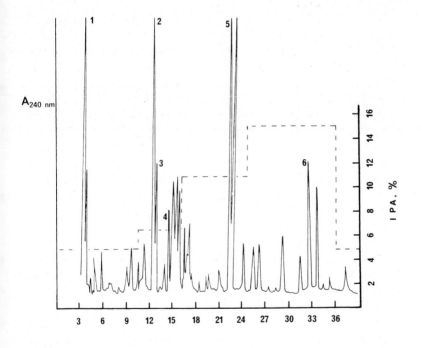

Fig. 11.8.2. HPLC separation of carotenoids in tobacco. Chromatographic conditions:
stationary phase, Spherisorb 3 μm silica (250×4.6 mm I.D.); mobile phase, 5%–14%
isopropanol (IPA) in n-hexane; flow rate, 1 ml/min; temperature, ambient; detection,
UV at 450 nm. Peaks: 1, carotenes; 2, lutein; 3, zeaxanthine; 4, lutein 5,6-expoxide; 5,
violaxanthine; 6, neoxanthine. Reproduced from Stewart et al. (1983), with permission.

of vitamins D_2, D_3 and E from cod liver oil using a reversed phase system (Stancher and Zonta, 1983). Major contaminants in the analysis of vitamins D_2 and D_3 are their respective metabolic precursors 7-dehydrocholesterol and ergosterol; however, both normal phase and reversed phase systems can achieve satisfactory separations. Detection of these vitamins is commonly carried out using UV detection at 254 nm.

11.8.2.3. E vitamins

HPLC is now accepted as the method of choice for the analysis of the E vitamins (the tocopherols) from plant tissue extracts. However, difficulties in the extraction of the E vitamins do not always leave the samples in a convenient chromatographic buffer and consequently other methods of analysis have also been used (Desai, 1984). Vitamin E activity resides in a number of chemical species which are generally classified as tocopherols and tocotrienols. The most common forms are α-, β-, γ- and δ-tocopherols and separation of each can be performed using either normal phase or reversed phase chromatography.

Analysis of tissue extracts containing vitamin E and its derivatives usually involves saponification followed by extraction into organic solvents.

A recent report recommends the use of acetone extraction followed by reversed phase chromatography for the analysis of tissue and plasma α-tocopherol levels (Zaspel and Saari-Csallany, 1983). The stationary phase was Lichrosorb RP-18 with a mobile phase of 2% water in methanol. An alternative system using a normal phase Micropak Si-5 support has also been reported to give reproducible estimations of tissue extracts of the tocopherols with a sensitivity of detection as low as 0.5–1.0 ng (Buthriss and Diplock, 1984).

Detection of α-tocopherol is best carried out using fluorescence detection at excitation and emission wavelengths of 295 nm and 340 nm, respectively. UV detection at 294 nm is also possible but less sensitive.

11.8.2.4. K vitamins

Extensive development of a two-stage chromatographic system for the
analysis of the K vitamins has been reported (Lefevre et al., 1979).
This method, which involves initial normal phase chromatography
followed by reversed phase chromatography, was necessary because
only small quantities of vitamin K are present in the extracts of milk

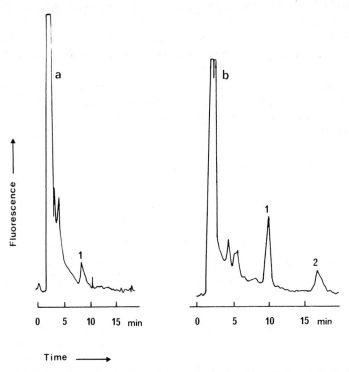

Fig. 11.8.3. Comparison of serum from a normal volunteer (a) and after oral adminis-
tration of 10 mg of phylloquinone (b). Chromatographic conditions: stationary phase,
Hypersil-MOS (5 μm) (100×3 mm I.D.); mobile phase, 7.5% water in methanol;
temperature, ambient; flow rate 0.9 ml/min; detection, post-column fluorescence
excitation at 320 nm, emission at 420 nm. Peaks: 1, phylloquinone (K1); 2, phylloquinone K2(30). Reproduced from Langenberg and Tjaden (1984), with permission.

and plasma. Alternative methodologies have been reported for plant extracts (Shearer, 1983). The use of the twin-column system described above has disadvantages for routine rapid analysis and a simplified system has been proposed in which vitamin K can be extracted into hexane and then directly subjected either to reversed phase or normal phase chromatographic analysis. In this study normal phase was found to be superior to reversed phase chromatography by simplifying the chromatographic profile (Wilson and Park, 1983). The samples were applied to a silica stationary phase and eluted under isocratic conditions with a mobile phase of 0.2% acetonitrile in hexane. Detection of compounds from the vitamin K group can be performed using UV at 248–254 nm. The sensitivity of detection has been reported to be at the subnanogram level in plasma extracts using a post-column colourimetric reduction system combined with either fluorescence or amperometric detection. Fluorescence detection has a minimum reported level of 25 pg and can be used in combination with either normal phase or reversed phase HPLC (Langenberg and Tjaden, 1984) (Fig. 11.8.3).

11.8.3. Water-soluble vitamins

11.8.3.1. Introduction

The HPLC analysis of water-soluble vitamins has been used extensively for over a decade and only the more recent developments will be discussed here. Comprehensive reviews of the early developments in the field have been written (Wittina and Hanley, 1976). The earliest separations were carried out using ion-exchange chromatography, reflecting the more polar nature of the water-soluble vitamins. However, the introduction of reversed phase ion-pair chromatography using such reagents as sulphonic acids, trifluoroacetic acid and inorganic phosphate has improved many of the earlier methods.

11.8.3.2. B_1 vitamin

Vitamin B_1 generally exists in either the pyrophosphate or the ester form. Before analysis, the ester may be enzymatically converted to the corresponding acid using acid phosphatase and the total vitamin B_1 content determined as a single species. Chromatographic separation of vitamin B_1 in whole blood, cerebrospinal fluid and milk can be performed by reversed phase ion-pair chromatography using a mobile phase of 0.05 M sodium citrate buffer, pH 4.0–methanol (55 : 45) plus 0.01 M sodium octane–sulphonate in combination with a stationary

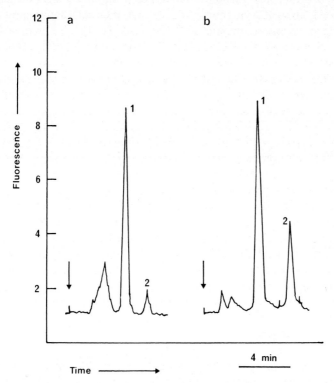

Fig. 11.8.4. HPLC of thiamine in human serum (a) and human cerebrospinal fluid (b). Chromatographic conditions: stationary phase, μBondapak C18 (10 μm) reversed phase (300 × 3.9 mm I.D.); mobile phase, methanol–aqueous sodium citrate, pH 4.0, 0.05 mol/l (45/55, v/v), sodium 1-octanesulphonate 10 mmol/l; temperature, ambient; flow rate, 1.2 ml/min; detection, post-column fluorescence (excitation at 367 nm, emission at 435 nm). Peaks: 1, salicylamide; 2, thiamine. Reproduced from Wielders and Mink (1983), with permission.

phase of C_{18} modified silica (Wielders and Mink, 1983) (Fig. 11.8.4). In this example, post-column derivatisation was carried out using potassium ferricyanide to oxidise thiamine to thiochrome which was then detected fluorimetrically (excitation at 367 nm and emission at 435 nm).

The determination of vitamin B_1 in multivitamin preparations can be hampered by extraction difficulties but the use of sodium diethylene penta-acetic acid facilitates recoveries in excess of 90% (Walker et al., 1981). Subsequently, reversed phase ion-pair chromatography can be carried out using a mobile phase of 0.001 M hexanesulphonic acid in 1% acetic acid–methanol (75 : 25). Using this method, thiamine, niacinamide, riboflavin and pyridoxine can be quantitated.

Normal phase chromatography using a Lichrosorb-NH_2 stationary phase in combination with a post-column derivatisation procedure has been used for the resolution and detection of thiamine and its derivatives. Using a mobile phase of acetonitrile–90 mM potassium phosphate, pH 8.4 (60 : 40), good separations of thiochrome, thiochrome monophosphate, pyrophosphate and triphosphate were obtained (Ishii et al., 1979). The sensitivity of detection using mixtures of the thiochrome derivatives was approximately 1 pmol.

The Shodex OH Pak M-414 normal phase support has been used to determine levels of thiamine in whole blood by using the ferricyanide post-column detection system described earlier. Samples were converted enzymatically to free thiamine prior to chromatography which was carried out with a mobile phase consisting of 0.2 M Na_2HPO_4 (Kimura et al., 1982). Only 0.1 ml of blood was required for the analysis.

11.8.3.3. B_2 vitamins

The most common forms of vitamin B_2 are riboflavin 5′-phosphate (FMN) and flavin adenine dinucleotide (FAD), which are best known for their participation as co-factors (ligands) to some of the enzymes involved in electron transfer chains. The ligand is usually coupled to enzymes through the phosphate moiety and therefore isolation from tissue can be achieved by either mild acid hydrolysis or by enzymatic digestion with an acid phosphatase.

The most popular chromatographic modes for the resolution of the flavins are reversed phase and normal phase chromatography. For example, flavins can be determined in meat extracts by using a stationary phase of silica gel with a mobile phase of chloroform–methanol (90 : 10) (Ang and Moseley, 1980). Identification of the by-products of FMN formed during chemical synthesis has recently been improved by the development of a reversed phase system using a C_{18} stationary phase and a mobile phase of methanol–0.1 M ammonium formate, pH 3.7 (17 : 83). Interestingly, alteration of the mobile phase to methanol–5 mM tetrabutylammonium formate, pH 3.5 (27 : 73) results in the elution order of the mono- and diphosphate forms being reversed (Nielsen et al., 1983).

Since FMN can be chromatographed on DEAE–Sephadex the newer hydrophilic ion-exchangers (TSK DEAE or Mono Q) may offer some advantages. FMN may be purified by affinity chromatography using apoflavodoxin immobilised on conventional supports. The intrinsic absorption of the various species of vitamin B_2 facilitates detection by monitoring UV absorbance at 280 nm; however, more commonly fluorescence detection is used (excitation at 450 nm and emission at 520 nm).

11.8.3.4. B_6 vitamins

Vitamin B_6 exists as six separate forms in the pyridine group of water-soluble vitamins. The common forms are pyridoxal and pyridoxamine together with their corresponding phosphate esters and pyridoxine forms. These compounds function as cofactors in a wide variety of enzyme reactions, but most notably in the transamination reaction of amino acid biosynthetic pathways. Extraction of this group of vitamins can be performed by the same methods as those described for the B_2 vitamins (Section 11.8.3.3).

Reversed phase systems have been used to separate the different members of the vitamin B_6 group and typical conditions employ a C_{18} stationary phase with a mobile phase of acetonitrile–aqueous phosphate buffer (Lim et al., 1980). Other popular methods are based on ion-exchange chromatography and a complete separation of pyridoxamine, pyridoxal, pyridoxine and their phosphate esters has been accomplished using this mode. In order to simplify the chromatography and to specifically detect pyridoxal phosphate a column switching device was used (Vanderslice and Maire, 1980). This method has since been used to analyse vitamin B_6 from a variety of sources including blood and food products. Detection of the B_6 vitamins can be performed by monitoring UV absorbance at 210–250 nm. Alternatively, fluorescence detection may be used for pyridoxamine phosphate, pyridoxine phosphate, pyridoxamine and pyridoxine at excitation and emission wavelengths of 310 nm and 380 nm, respectively. Pyridoxal and pyridoxal phosphate may also be detected by using fluorescence detection (excitation at 280 nm and emission at 487 nm) if they are initially converted to their oxime derivatives by the inclusion of semicarbazide in the mobile phase (Vanderslice and Maire, 1980).

11.8.3.5. Folic acid

Folic acid is the parent compound of a complex mixture of conjugated pterins which function in the transfer of 2-carbon groups in a number of biosynthetic pathways. Folates are particularly sensitive to oxidation by heat, UV light or by exposure to acid (Baugh and Krundieck, 1971). Ion-exchange separations of folates have been reported on pellicular supports (Reed and Archer, 1976), although only a few of the naturally occurring folates were included in these studies.

A method has been described which allows the analysis of all the forms of folic acid in tissue extracts (Dutch et al., 1983). The polyglutamate forms were enzymatically hydrolysed with rat liver conjugase to produce the monoglutamate derivatives which were then subjected to ion-exchange chromatography on Dowex 50 X-4. Samples were eluted from the column by a step-wise increase in pH and then lyophilised. This chromatographic procedure eliminates contaminating UV-absorbing material and allows subsequent separation and detection of the individual folates using reversed phase ion-pair chromatography. The latter was carried out on a μBondapak C_{18} column using a mobile phase of methanol–0.01 M ammonium dihydrogen phosphate/0.005 M tetrabutylammonium phosphate (20 : 80)

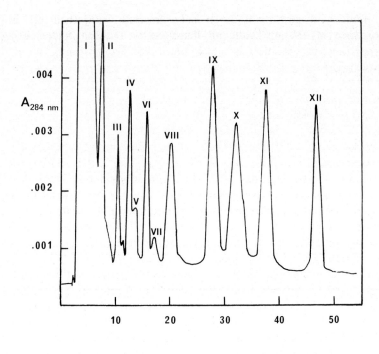

Fig. 11.8.5. HPLC analysis of folate cofactors. Chromatographic conditions: stationary phase, μBondapak C18 (300$\times$3.9 mm I.D.); mobile phase, 20% methanol–0.01 M ammonium dihydrogen phosphate–5 mM tetrabutylammonium phosphate; flow rate, 1 ml/min; temperature, ambient; detection, UV at 284 nm. Peaks: I, ascorbate, 5,10-CH=H_4PteGl7; II, mercaptoethanol; III, PABGA (2.5 ng); IV, 10-CHO-H_4PteGlu (28 ng); V, 10-CHO-H_2PteGlu; VI, H_4PteGlu (18.8 ng); VII, 10-CHO-PteGlu; VIII, 5-CHO-H_4PteGlu (12.5 ng); IX, H_2PteGlu (12.5 ng); X, PteGlu (9.4 ng); XI, 5-CH_3-H_4PteGlu (25 μg); XII, 10-CH_3PteGlu. Reproduced from Dutch et al. (1983), with permission.

(Fig. 11.8.5). The eluting folates were detected by monitoring their UV absorbance at 284 nm. Reversed phase chromatographic systems have been used to detect folates in serum (Birmingham and Green, 1983) and in multivitamin preparations (Holcombe and Fusari, 1981).

Folic acid derivatives are usually detected using their UV absorbance at 280 nm although detection of reduced folates using amperometric detection has also been reported (Birmingham and Green, 1983).

11.8.3.6. Ascorbic acid

Ascorbic acid (vitamin C) is labile to oxidation and is readily converted to dehydroascorbic acid which can subsequently break down into biologically inactive species. Oxidation may be avoided during extraction and chromatography by maintaining a moderately low pH environment; this makes mobile phase conditions ideally suited to chromatography on modified silica supports and several methods have been reported, including reversed phase (Tsao and Salin, 1981) and ion-exchange (Lieber et al., 1981). Normal phase HPLC has also been used to separate L-ascorbic acid from D-ascorbic acid using a Lichrosorb NH$_2$ column in combination with a mobile phase of acetonitrile–0.005 M potassium dihydrogen phosphate, pH 4.4–4.7 (75 : 25) (Bui-Nguyen, 1980). However, because of the versatility of reversed phase ion-pair chromatography, the majority of analyses can be performed by this method alone. Typical ion-pair chromatographic conditions utilise a mobile phase of 0.01–0.1 M acetate, pH 4–6 to which an organic modifier such as ethanol or methanol has been added. The most popular stationary phases are C$_8$ and C$_{18}$ bonded silica. Both electrochemical and UV detection (264 nm) methods have been used; however, an alternative system with greater sensitivity utilises an initial enzymatic conversion of L-ascorbate to L-dehydroascorbic acid which is then chromatographically resolved by reversed phase chromatography (Speek et al., 1984). A post-column derivatisation system can then used in which L-dehydroascorbic acid is reacted with o-phenylenediamine to form its quinoxaline derivative. This derivative can be monitored by fluorescence detection (excitation at 365 nm and emission at 418 nm). This methodology facilitates the detection of ascorbic acid at levels below those found in blood samples.

11.8.3.7. Nicotinic acid

Nicotinic acid was first identified as an essential dietary supplement in 1937. It can be found in an active form as either free nicotinic acid, the amide derivative or bound to the pyridine-containing nucleotides (nicotinamide adenine dinucleotide (NAD) and the corresponding phosphate ester (NADP)).

Nicotinic acid and its derivatives are stable to oxidation by heat, light, acid or alkali and this means that extraction into solvent systems compatible with HPLC is relatively easy by comparison with other vitamins. Biological extracts are readily prepared by deproteinisation with acetone followed by extraction with dilute hydrochloric acid; alternatively, ethyl acetate in combination with hydrochloric acid may be used to extract samples. The most popular HPLC mode for the separation of nicotinic acid is reversed phase ion-pair chromatography. For example, using a μBondapak C_{18} column with a mobile phase of water–methanol (9 : 1) plus 0.05 M tetrabutylammonium phosphate as the ion-pair reagent, nicotinamide-N-oxide, 2-hydroxypyridine-5-carboxylic acid, nicotinamide, nicotinic acid and nicotinuric acid were consecutively eluted (Hengen et al., 1978).

The main problem encountered when monitoring nicotinic acid derivatives which have been extracted from biological samples is the low level of sensitivity achieved using UV detection at 254 nm. However, using a reversed phase HPLC system in combination with a series of three UV detectors, each monitoring different wavelengths, approximately seventy metabolically active organic acids, including nicotinic acid, could be identified by comparing retention time and peak height ratios (Todoriki et al., 1983).

Several methods of pre-column derivatisation have been investigated to improve sensitivity of detection and early reports suggested that the use of cyanogen bromide coupled with *p*-aminoacetophenone provided a sensitive fluorescence detection system (excitation at 435 nm and emission at 500 nm). However, a recent report presents a simple alternative using derivatisation with N, N'-dicyclohexyl-O-(7-methoxycoumarin-4-yl) methylisourea in which nicotinic acid concentrations as low as 0.2 nmol/ml were detectable in serum samples of only 100 μl (Tsuruta et al., 1984). The nicotinic acid derivatives

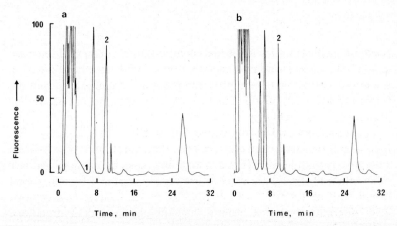

Fig. 11.8.6. HPLC determination of nicotinic acid in (a) normal serum and (b) serum fortified with nicotinic acid (50 nmol/ml). Chromatographic conditions: stationary phase, Lichrosorb RP-18 (Merck) (5 μm) (150×4 mm I.D.); mobile phase, acetonitrile–water (4:6 v/v) containing 5 mM sodium 1-hexane sulphonate; flow rate, 1.3 ml/min; temperature, ambient; detection, pre-column derivatisation, excitation at 325 and 335 nm and emission at 395 and 390 nm, respectively. 1, nicotinic acid; 2, 2-chloronicotinic acid. Reproduced from Tsuruta et al. (1984), with permission.

were separated using a reversed phase ion-pair HPLC system with a mobile phase of acetonitrile–water (4:6) containing 5 mM sodium 1-hexanesulphonate on a Lichrosorb C_{18} stationary phase (Fig. 11.8.6).

11.8.3.8. B_{12} vitamin

The importance of vitamin B_{12} determinations in the diagnosis of various disease states, together with the recognition that a large number of its analogues are biologically active, has resulted in a great deal of effort being directed to developing reliable quantitative assays. Extraction of vitamin B_{12} and its derivatives can be carried out using organic solvents; for example, a mixture of ether–acetone (3:1) may be used (Freukel et al., 1980).

Both analytical and preparative separations have been carried out using normal phase conditions on a pellicular silica support (Snyder and Kirkland, 1979). Most analyses are currently performed using reversed phase HPLC, which was first applied to vitamin B_{12} in 1974 (Stillman and Ma, 1974). More recently, the resolution and quantitation of cobalamin coenzymes has been reported using a C_{18} stationary phase in combination with a mobile phase of 50 mM sodium phosphate–methanol (Freukel et al., 1980).

Intense interest in the metabolism of vitamin B_{12} analogues in microorganisms has led to the development of alternative HPLC systems, although most are based on reversed phase HPLC. In the analysis of cobalamin coenzymes and corrinoids from *Lactobacillus leichmannii*, extracts were prepared by warming the cells to 70 °C in 80% ethanol. After clarification, the extracts were dried by evaporation and redissolved in water. A simple isocratic system of acetonitrile–1 mM ammonium acetate, pH 4.4 (3 : 7) with either a C_8 or C_{18} stationary phase was sufficient to separate the major naturally occurring cobalamins (sulfitocobalamin, cyanocobalamin, methylcobalamin and aquacobalamin). However, a mobile phase gradient system (buffer A, 0.05 M phosphoric acid, pH 3.0; buffer B, acetonitrile) on a C_8 or C_{18} stationary phase gave improved resolution along with five other aminoalkylcobalamin homologues (Jacobsen et al., 1982). Cobalamins from murine leukaemic cells were also analysed in this study.

Gradient reversed phase HPLC has also been used in the analysis of cobamides present in *Methanobacterium bryantii* (Whitman and Wolfe, 1984). After initial purification by isocratic reversed phase HPLC at neutral pH the extracts were subjected to gradient HPLC using a C_{18} stationary phase with an initial mobile phase of 100 mM lithium chloride–methanol (76 : 24) and a final mobile phase of 100 mM lithium chloride–methanol (52 : 48). Detection by UV absorbance at 254 nm allowed quantitation of as little as 1 pmol of cyanocobalamin.

11.8.3.9. Biotin

Current assays for the detection and analysis of biotin and its analogues either rely on enzyme-linked assays or utilise the high affinity of biotin for avidin (Silver, 1979).

Early development of chromatographic analyses for biotin used open column chromatography with ion-exchange supports (Salib et al., 1979). This technique is extremely time consuming and in addition detection of the eluted products is made difficult because of the poor UV absorbance of biotin. These problems have resulted in the development of pre-column derivatisation procedures combined with gradient HPLC techniques.

A relatively simple method for the analysis of biotin and its analogues has been devised using either 2,4'-dibromoacetophenone (DBAB) or 4-bromomethyl methoxycoumarin (MCM) as derivatisation agents to facilitate fluorescence detection (excitation at 360 nm and emission at 410 nm) (Desbene et al., 1983). After derivatisation, the samples were first chromatographed on silica to remove by-products of the derivatisation procedure and were then applied to a C_{18} stationary phase. Elution was achieved using a gradient of tetrahydrofuran–water (30 : 70 to 60 : 40 or 40 : 60). Derivatisation with DBAB and UV detection at 254 nm has been reported to give a detection limit of approximately 0.5 pmol/μl. Using MCM derivatives and fluorescence detection the limit of sensitivity was reported to be increased by a factor of ten. An equally high sensitivity has been reported for DBAB esters of biotin using an HPLC system coupled to a mass spectrometer (Azoulay et al., 1984).

11.8.3.10. Lipoic acid

Lipoic acid can exist either free or bound to protein through the epsilon amino group of lysine residues. More than 90% of the bound material can be recovered by treatment with 6 M HCl at 110°C for 24 h. An extensive literature exists describing systems for the analysis of lipoic acid and its derivatives by thin layer chromatography (Bayer et al., 1979), but reversed phase HPLC is at present the method of choice for the separation of lipoic acid and its analogues. Using a C_{18} stationary phase and a mobile phase gradient from 25% to 75% methanol in acetic acid, eleven lipoic acid derivatives were resolved, including: lipoic acid, dihydrolipoic acid, lipoamide, lipoyl glycinamide, 8-methyl-lipoic acid, 6,9-dithiononanoic acid, methyl tetranorlipoate S-oxide, tetranorlipoic acid, β-hydroxybisnorlipoic acid, bisnorlipoic acid and methyl lipoate (Howard and McCormick, 1981).

Elution was monitored by UV absorbance at either 330 nm, 240 nm or 285 nm depending on the derivative.

11.8.3.11. Pantothenic acid

There are a variety of methods available for the separation of pantothenate derivatives including thin layer chromatography and gas–liquid chromatography. Several modes of HPLC have been used for the separation of pantothenates, including reversed phase, normal phase and ion-exchange. Reversed phase HPLC provides the greatest flexibility for the separation of pantothenate intermediates and, using a stationary phase of C_{18} modified silica with a series of phosphate–methanol buffers, it was possible to separate CoASH, dephospho-CoA, pantothenic acid, 4'-pantothenic acid, 4-phosphopantothenoyl-1-cysteine and 4'-phosphopantetheine (Halvorsen and Skrede, 1980). An alternative reversed phase system using a C_{18} stationary phase combined with a mobile phase of 0.9 M acetic acid has been used for the analysis of calcium pantothenate concentrations in multivitamin preparations (Jonvel et al., 1983).

Ion-exchange HPLC has been used for the quantitation of CoASH in biological samples using a strong anion-exchange support (Partisil SAX) as the stationary phase and a mobile phase of 196 mM potassium phosphate, pH 3.9–isopropanol–thiodiglycol (98 : 2 : 0.05) (Ingebretsen et al., 1979). Picomole levels of CoASH could be quantitated using UV detection at 254 nm.

11.8.4. References

Vitamin	Source	Chromato-graphic mode	Reference
A vitamins: retinoids	Serum	RPC	Annesley, T., Grachero, D., Wilherson, K., Grekin, J. and Ellis, C. (1984) J. Chromatogr. 305, 199–203.
	Milk	RPC	Barnett, S.A., Frick. L.W. and Baine, H.J. (1980) Anal. Chem. 52, 610–614.

Vitamin	Source	Chromato-graphic mode	Reference
	Foods	RPC	Egberg, D.C., Heroff, J.C. and Potter, R.W. (1977) J. Agric. Food Chem. 25, 1127–1132.
	Milk, margarine	NPC	Thompson, J.N., Hatina, G., Maxwell, W.B. (1980) J. Ass. Off. Anal. Chem. 63, 894–898.
	Cereal	NPC	Widicus, W.A. and Kirk, J.R. (1979) J. Ass. Off. Anal. Chem. 62, 636–643.
A vitamins: carotenes	Tobacco	RPC	Bevan, J.L., Bailey, P.A. and Stewart, P.S. (1983) J. Chromatogr. 782, 589–593.
	Spinach	RPC	Braumann, T. and Grimme, C.H. (1981) Biochim. Biophys. Acta 637, 8–12.
	Free	RPC	Krinsky, N.I. and Wecankiwar, S. (1984) Methods Enzymol. 105, 155–159.
	Cheese	NPC	Stancher, B. and Zonta, F. (1983) J. Chromatogr. 286, 93–98.
	Margarine	NPC	Thompson, J.N., Hatina, G. and Maxwell, W.B. (1980) J. Ass. Off. Anal. Chem. 63, 894–898.
	Tomato	RPC	Zakaria, M., Simpson, K., Brown, P.R. and Krstulovic, A. (1979) J. Chromatogr. 176, 109–117.
D vitamins	Fish	NPC	Egars, E. and Lamberstein, G. (1979) Int. J. Vitam. Nutr. Res. 49, 35–43.
	Milk	NPC	Cohen, H. and Wakeford, B. (1980) J. Ass. Off. Anal. Chem. 63, 1163–1167.
	Milk	RPC	Henderson, S.K. and McLean, L.A. (1978) J. Ass. Off. Anal. Chem. 62, 1358–1360.
	Cod liver oil	RPC	Stancher, B. and Zonta, F. (1983) J. Chromatogr. 286, 93–98.
	Milk	NPC	Thompson, J.N., Maxwell, W.B. and L'Abbe, M. (1977) J. Ass. Off. Anal. Chem. 60, 998–1002.
E vitamins	Milk	RPC	Barnett, S.A., Frick, L.W. and Baine, H.M. (1980) Anal. Chem. 52, 610–614.

Vitamin	Source	Chromato-graphic mode	Reference
	Tissue	NPC	Buthriss, J.L. and Diplock, A.T. (1984) Methods Enzymol. *105*, 131–138.
	Tissue	NPC	De Leenheer, A.P., De Bevere, V.D., Cruyl, A.A. and Claeys, A.E. (1978) Clin. Chem. *24*, 585–590.
	Foods	RPC	Devries, J.W., Egbert, D.C. and Heroff, J.C. (1979) In: Liquid Chromatographic Analysis of Food and Beverages, Vol. 2. Academic Press, New York, NY.
	Blood	NPC	Hatfim, L.J. and Kaysen, H.J. (1979) J. Lipid Res. *20*, 639–645.
	Vegetable oil, cereals, milk	NPC	Thompson, J.N. and Hatina, G. (1979) J. Chromatogr. *2*, 327–344.
	Rat brain	NPC	Vatassery, G.J., Maynard, V.R. and Hagen, D.F. (1978) J. Chromatogr. *161*, 299–302.
	Tissue	RPC	Zaspel, B.I. and Saari-Csallany (1983) Anal. Biochem. *130*, 146–150.
K vitamins	Dairy products	RPC	Barnett, S.A., Frick, L.W. and Baine, H.M. (1980) Anal. Chem. *52*, 610–614.
	Plasma	NPC/RPC	Langenberg, J.P. and Tjaden, V.R. (1984) J. Chromatogr. *305*, 61–72.
	Milk and plasma	NPC/RPC	Lefevre, M.P., De Leenheer, A.P. and Claeys, A.E. (1979) J. Chromatogr. *186*, 749–762.
	Peanuts	NPC/RPC	Shearer, M.J. (1983) Adv. Chromatogr. *21*, 243–301.
	Foods	NPC	Thompson, J.N. and Hatina, G. (1979) J. Liq. Chromatogr. *2*, 327–344.
	Plasma	NPC/RPC	Wilson, A.E. and Park, B.K. (1983) J. Chromatogr. *277*, 292–299.
B₁ vitamins	Meat	NPC	Ang, C.Y. and Moseley, F.A. (1980) J. Agric. Food Chem. *28*, 483–486.
	Free thiocromes	NPC	Ishii, K., Sarai, K., Sanemori, H. and Kawasaki, T. (1979) Anal. Biochem. *97*, 191–195.
	Cereal	RPC	Kammai, J.F., Labuza, T.P. and Warthulsen, J.J. (1980) J. Fol. Sci. *45*, 1497–1501.

Vitamin	Source	Chromato-graphic mode	Reference
	Plasma	NPC	Kimura, M., Fujita, T. and Itokawa, Y. (1982) Clin. Chem. 28, 29–31.
	Rice	RPC	Toma, R.B. and Tabekhia, M.M. (1979) J. Fol. Sci. 44, 263.
	Multivitamins	RPC	Walker, M.C., Carpenter, B.E. and Cooper, E.L. (1981) J. Pharm. Sci. 70, 99–101.
	Milk, blood and cerebrospinal fluid	RPC	Wielders, J.P.M. and Mink, C.J.K. (1983) J. Chromatogr. 277, 145–156.
B₂ vitamins	Meat	NPC	Ang, C.Y.W. and Morseley, F.A. (1980) J. Agric. Food Chem. 28, 483–486.
	Chemical synthe-sis	RPC	Nielsen, P., Rauschenbach, R. and Bacher, A. (1983) Anal. Biochem. 130, 359–368.
	Orange juice	NPC	Roussel, R. (1979) In: Liquid Chromatographic Analysis of Food and Beverages. Academic Press, New York NY. Vol. 1, p. 161.
B₆ vitamins	Foods	RPC	Gregory, J.F. (1980) J. Agric. Food Chem. 28, 486–489.
	Milk	RPC	Lim, K.L., Young, R.W. and Driskell, J.A. (1980) J. Chromatogr. 188, 285–288.
	Plasma	IEX	Vanderslice, J.T. and Maire, C.E. (1980) J. Chromatogr. 196, 176–179.
	Fruit/vegetables	IEX	Wong, F.F. (1978) J. Agric. Food Chem. 26, 1444–1448.
Folic acid	Serum	RPC	Birmingham, B.K. and Green, D.S. (1983) J. Pharm. Sci. 72, 1306–1309.
	Tissue	IEX/RPC	Dutch, D.S., Bowers, S.W. and Nichol, C.A. (1983) Anal. Bio-chem. 130, 385–392.
	Multivitamins	RPC	Holcombe, I.J. and Fusari, S.A. (1981) Anal. Chem. 53, 607–609.
	Multivitamins	RPC	Horne, D.W., Briggs, W.T. and Wagner, C. (1981) Anal. Biochem. 116, 393–397.
	Free derivatives	IEX	Reed, L.S. and Archer, M.C. (1976) J. Chromatogr. 121, 100–103.

Vitamin	Source	Chromato-graphic mode	Reference
Vitamin C	Potato	RPC	Augustin, J., Beck, C.B. and Marousek, G.A. (1981) J. Fol. Sci. *46*, 312–316.
	Lymphocytes	IEX	Lieber, L.F., Kai, S., Krigel, R., Pelle, E. and Silber, R. (1981) Anal. Biochem. *118*, 53–57.
	Fruit juice	RPC	Pachla, L.A. and Kissinger, P.T. (1979) Methods Enzymol. *62*, 15–24.
	Plasma	RPC	Tsao, C.S. and Salin, S.L. (1980) J. Chromatogr. *224*, 477–480.
	Plasma	RPC	Speek, A.J., Schriver, J. and Schreurs, W.H.P. (1984) J. Chromatogr. *305*, 53–60.
Nicotinic acid	Free	NPC	Bui-Nguyen, M.H. (1980) J. Chromatogr. *196*, 163–165.
	Serum	RPC	Hengen, N., Seiberth, V. and Hengen, M. (1978) Clin. Chem. *24*, 1740–1743.
	Free	RPC	Lingeman, H., Hulsoff, A., Underberg, W.J.M. and Offerman, F.B.J.M. (1984) J. Chromatogr. *290*, 215–222.
	Urine	RPC	Todoriki, H., Hayashi, T. and Naruse, H. (1984) J. Chromatogr. *310*, 273–301.
	Serum	RPC	Tsuruta, Y., Kohashi, K. Ishida, S. and Onkura, Y. (1984) J. Chromatogr. *309*, 309–315.
	Cereal	RPC	Tyler, T.A. and Schrago, R.A. (1980) J. Liq. Chromatogr. *5*, 269–275.
	Free	RPC	Walker, M.C., Carpenter, B.E. and Cooper, E.L. (1981) J. Pharm. Sci. *70*, 1–7.
B₁₂ vitamins	Micro-organism	RPC	Beck, R.A. and Brink, J.J. (1976) Environ. Sci. Technol. *10*, 173–175.
	Various	RPC	Freukel, E., Prough, R. and Kitchens, R.L. (1980) Method Enzymol. *67*, 31–40.
	Various	RPC	Jacobsen, D.W., Green, R., Quadros, E.V. and Montejano, V.D. (1982) Anal. Biochem. *120*, 394–403.

Vitamin	Source	Chromatographic mode	Reference
	Free	NPC	Snyder, L.R. and Kirkland, J.J. (1979) Introduction to Modern Liquid Chromatography, 2nd edn. J. Wiley, New York, NY.
	Micro-organism	RPC	Stillman, R. and Ma, S. (1974) Microchim. Acta. 82, 641–648.
	Micro-organism	RPC	Whitman, W.B., and Wolfe, R.S. (1984) Anal. Biochem. *137*, 261–265.
Biotin	Free/micro-organism	RPC	Azoulay, S., Desbene, P.L., Frappier, F. and Georges, Y. (1984) J. Chromatogr. *303*, 272–276.
	Free	RPC	Desbene, P.L., Coustal, S. and Frappier, F. (1983) Anal. Biochem. *128*, 359–362.
	Micro-organism	IEX	Salib, A.G., Frappier, F., Guillerm, G. and Marquet, A. (1979) Biochem. Biophys. Res. Commun. *88*, 312–319.
Lipoic acid	Free	RPC	Howard, S.C. and McCormick, D.B. (1981) J. Chromatogr. *208*, 129–131.
Pantothenic acid	Free	RPC	Halvorsen, O. and Skrede, S. (1980) Anal. Biochem. *107*, 103–108.
	Multivitamin	RPC	Hudson, T.J. and Allen, R.J. (1984) J. Pharm. Sci. *73*, 111–113.
	Free	IEX	Ingerbretsen, O.C., Normann, D.T. and Flatmark, T. (1979) Anal. Biochem. *96*, 181–188.
	Multivitamin	RPC	Jonvel, P., Andermann, G. and Bartholemy, J.F. (1983) J. Chromatogr. *281*, 371–376.

Antibiotics

An antibiotic is defined as a chemical compound which displays growth inhibitory activity against micro-organisms. One of the earliest and best known antibiotics is penicillin, which was isolated as a product from a microbial fermentation; subsequently, a variety of chemical types of antibiotics with a wide spectrum of activity have been discovered. Moreover, the ability of the synthetic chemist to modify such molecules has led to an equally large number of semisynthetic antibiotics. Many micro-organisms have mutated to overcome the inhibitory action of antibiotics by producing enzymes which are able to degrade them into harmless compounds (e.g. β-lactamase in *E. coli* is able to break down the β-lactam ring in the penicillins). The emergence of antibiotic resistance provides the stimulus for the pharmaceutical industry to continue research into the discovery of new and chemically different classes of antibiotics.

Many antibiotics exhibit toxic side-effects in vivo and their successful use in the clinical situation requires that they be accurately monitored in biological fluids. Similarly, accurate determinations of antibiotic production in microbial fermentation broths is a critical aspect of the manufacturing process. In the past, antibiotics were monitored in biological fluids by taking various dilutions and measuring their inhibitory action in vitro. Such assay techniques are time consuming and also do not allow quantitation of less active metabolic products. Alternative methods of detection such as spectrophotometric techniques, thin layer chromatography, gas–liquid chromatography and radioimmunoassay have been developed but each of these suffers from drawbacks such as the requirement for exhaustive extraction or derivatisation prior to their assay. Thus HPLC has become the method of choice for the analysis and quantitation of antibiotics. The diverse chemical nature of antibiotic compounds makes it difficult to generalise on specific HPLC techniques and it is therefore necessary to consider each class of compound separately. In this chapter only the main classes of antibiotics will be reviewed, with a limited number of examples drawn from the remainder. An excellent review containing preferred HPLC procedures for a large number of antibiotics can be found in the literature (Nilsson-Ehle, 1983; DeLuca and Papadimitriou, 1983).

The major classes of antibiotics may be subdivided into β-lactams, aminoglycosides, tetracyclines, peptides, anthracyclines and others.

11.9.1. General considerations

HPLC analysis of antibiotics in biological fluids does not require special sample preparation other than the techniques previously described (Chapter 10). Removal of protein by either acid precipitation or by ultrafiltration is satisfactory for serum samples. For urine samples an initial solvent extraction usually provides an adequate clean-up step.

Most antibiotics contain a chromophore which allows detection by UV absorption and by careful wavelength selection minimal interference from other components is encountered (Nilsson-Ehle et al., 1976). Fluorimetric detection is useful in monitoring aminoglycosides which show little UV absorption, although sample derivatisation is required for these compounds.

11.9.2. β-Lactams

This group of compounds includes both the penicillins and the cephalosporins for which the general chemical structures are shown in Fig. 11.9.1. Variation of R_1, R_2 and R_3 has allowed a number of clinically useful semisynthetic compounds to be synthesised. R_1, R_2

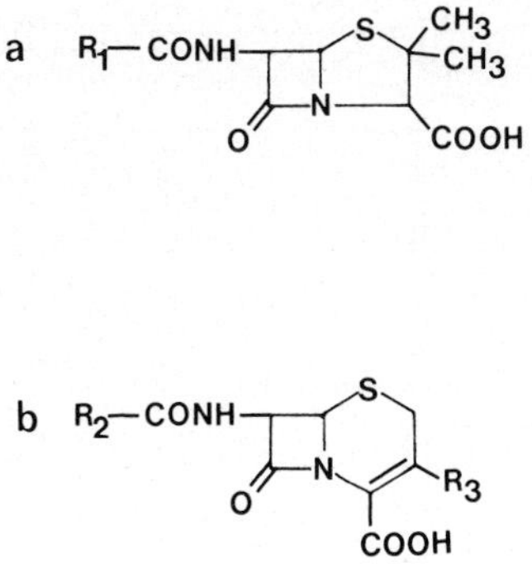

Fig. 11.9.1. The chemical structure of a penicillin (a) and a cephalosporin (b).

and R_3 are generally aromatic in nature. The carboxyl side-chain of these compounds usually means that these compounds exist as anions in neutral aqueous solution although where an additional amino group side-chain is present they may be amphoteric. Ion-exchange chromatography has been used for the separation of these compounds (Buhs et al., 1974) although, due to the usual lifetime limitations of such column packings, more recently reversed phase or reversed phase ion-pair chromatography has been preferred.

A variety of column packings and mobile phases have been used to resolve the penicillins (White et al., 1977; Lecaillon et al., 1982; Lebelle et al., 1980). A review by Miller and Neuss (1983) describes HPLC methods for the analysis of individual β-lactam antibiotics.

A standard technique for the HPLC analysis of amino penicillins (ampicillins) has been to use an aqueous acid phosphate buffer–acetonitrile mobile phase at a pH close to the isoelectric points of the amino penicillins (Margosis, 1982). Amino penicillins are most stable at their isoelectric point and despite their low solubility at this pH, it is optimal for their chromatography. The stationary phase used was bonded ODS on silica and elution was achieved under isocratic conditions (Fig. 11.9.2). Recently, microbore columns have been used in the separation of cephalosporin antibiotics (Fig. 11.9.3) and provide good baseline separations (White and Laufer, 1984).

Baseline resolution of standard mixtures of cephalosporins has

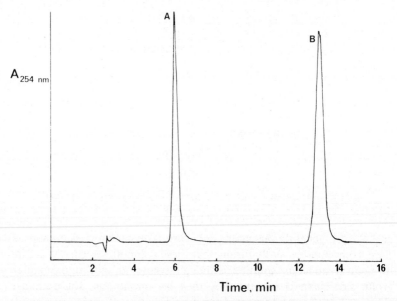

Fig. 11.9.2. Chromatogram of freshly dissolved ampicillin (A) and caffeine (B). Chromatographic conditions: column, Partisil PXS-525 ODS (250×4.6 mm I.D.); mobile phase, 0.01 M potassium dihydrogen phosphate and 0.001 M acetic acid (giving a pH of approximately 4.1) and containing 8% acetonitrile; flow rate, 1.5 ml/min; temperature, ambient; detection, UV at 254 nm. Reproduced from Margosis (1982), with permission.

been achieved using a mobile phase of acidic phosphate buffer–methanol in combination with a μBondapak C_{18} stationary phase. The levels of cephalosporin C levels in fermentation broths can be determined using these conditions.

Several of the metabolic precursors of β-lactams lack a suitable chromophore and derivatisation of their primary amino group with a fluorophore is necessary in order to allow their determination in fermentation broths. Dansyl chloride (Peng et al., 1977b), fluorescamine (Granneman and Sennello, 1982) and o-phthalaldehyde (Rogers et al., 1983; Crossley et al., 1980) have been used as derivatising agents. Derivatisation with o-phthalaldehyde requires reaction conditions of pH 9.0 (Peng et al., 1977b), a pH at which β-lactams are spontaneously hydrolysed and consequently post-column derivatisa-

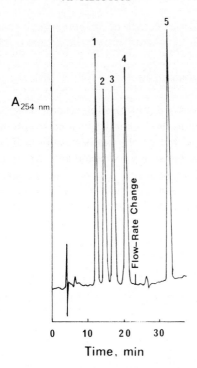

Fig. 11.9.3. Separation of a mixture of cephalosporin antibiotics on a microbore column. Chromatographic conditions: column, μBondapak C$_{18}$ (10 μm) (250×1.0 mm I.D.); mobile phase, 0.01 M sodium dihydrogen phosphate–methanol (75:25); flow rate, 50 μl/min for 23 min and 150 μl/min afterwards; detection, UV at 254 nm. Peaks: 1, Cephalexin; 2, Cefoxitin; 3, Cephadrine; 4, Cephaloglycin; 5, Cephalothin. Reproduced from White and Laufer (1984), with permission.

tion is necessary. The preferred conditions for the resolution of β-lactams employ reversed-phase ion-pair HPLC with a stationary phase of Spherisorb C$_8$ and a mobile phase of aqueous phosphate–acetonitrile–tetra-*n*-butylammonium hydroxide buffer.

The fluorescent detection of β-lactams in plasma has also been described and relies on the generation of specific naturally fluorescent degradation products (Miyazaki et al., 1983) which were initially chromatographed on a reversed phase column and then eluted with a mobile phase of methanol–water (3:2). The eluant was monitored using an excitation wavelength of 345 nm and an emission wavelength of 420 nm.

11.9.3. Aminoglycosides

The aminoglycosides are comprised of several subgroups including the gentamycins, kanamycins, neomycins and streptomycins. Generally the high polarity of these compounds results in limited retention on reversed phase systems and consequently reversed phase ion-pair chromatography coupled with a derivatisation step is used for the resolution of the aminoglycosides (Mays et al., 1976; Anhalt and Brown, 1978; Anhalt, 1977). Alternatively, the primary amino groups

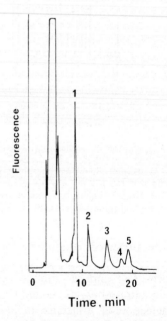

Fig. 11.9.4. Chromatograms of human serum spiked with tobramycin and gentamycin. Chromatographic conditions: column, Spherisorb 5-ODS (250 × 4.6 mm I.D.); mobile phase, methanol–water–0.2 M EDTA (80 : 15 : 5) pH 7.2; detection, fluorimetric (excitation at 340 nm and emission at 455 nm). Peaks: 1, tobramycin; 2–5, gentamycins. Reproduced from Marples and Oates (1982), with permission.

of the aminoglycosides can be derivatised with less polar aromatic groups to facilitate reversed phase chromatography (Peng et al., 1977a; Maitra et al., 1977; Wong et al., 1982).

Gentamycin is a widely used antibiotic but presents a number of undesirable side-effects when administered in high doses. Thus, it is vital to be able to monitor serum levels accurately. In the past this has been achieved using a number of assay methods other than HPLC. However, recently an HPLC method which compares favourably with microbiological assay methods has been described (Marples and Oates, 1982). This system utilises pre-column derivatisation with o-phthalaldehyde to facilitate fluorimetric detection with subsequent resolution on a Spherisorb 5-ODS reversed phase support using isocratic conditions with methanol–water–EDTA (Fig. 11.9.4). Baseline resolution of tobramycin and four gentamycin components was achieved by this method.

A further alternative utilises initial sample derivatisation with o-phthalaldehyde and subsequent separation on a cation-exchange support coupled to a reversed phase support. The mobile phase was a methanol–aqueous sodium acetate mixture the composition of which was altered for the specific compound being separated (Essers, 1984).

11.9.4. Tetracyclins

Tetracyclin is a derivative of four chemically fused ring systems whose general structure is shown in Fig. 11.9.5. Other tetracyclins such as 4-epitetracyclin, oxytetracyclin, chlorotetracyclin and anhydrotetracyclin are simple chemical derivatives of this structure and each may be easily resolved by reversed phase HPLC. The isocratic system in

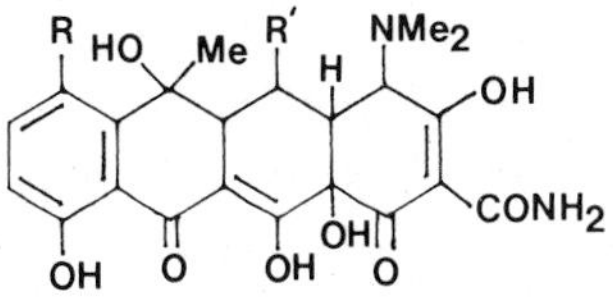

Fig. 11.9.5. Chemical representation of the general structure of tetracyclins.

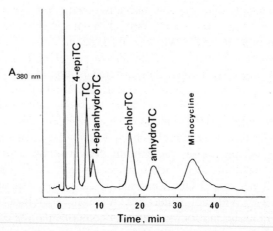

Fig. 11.9.6. Separation of tetracyclins (TC). Chromatographic conditions: column, Vydac RP (C_{18} ODS) (10 μm) (250×4.6 mm I.D.); mobile phase, 18% methanol in 0.001 M EDTA adjusted to pH 6.6; flow rate, 1.8 ml/min; detection, UV at 380 nm. Reproduced from Aszalos et al. (1982a), with permission.

Fig. 11.9.6 has proved to be particularly useful for the analysis of impurities in commercial preparations. It is known that tetracyclins epimerise at relatively low pH values and care must be taken not to produce artifactual peaks during chromatographic procedures.

11.9.5. Peptide antibiotics

Peptide antibiotics include a wide range of compounds such as cycloserine (an amino acid), bacitracin (a cyclic peptide), thiostrepton (a linear peptide), neocarzinostatin (a polypeptide) and bleomycin (a glycoprotein). These compounds possess either antibacterial and/or antitumor activity and should be regarded as conventional peptides for the purposes of HPLC. For a detailed overview of the HPLC retention times of the more common peptide antibiotics in various buffer systems the reader is referred to a recent review (Aszalos and Aquilar, 1984). The recommended stationary phase is a Vydac RP-18 column because it produced the least amount of peak tailing. Mobile phases contained either heptane sulphonic acid or phosphate buffer at various pH values and used methanol, acetonitrile or tetrahydrofuran as an organic modifier. Similar techniques have been described for the HPLC analysis and isolation of streptogramin (Grell et al., 1984) and vancomycin-type antibiotics (Sztarickskai et al., 1983).

11.9.6. Anthracyclins

Streptomyces cultures produce a group of antitumor antibiotics referred to as anthracyclins of which daunorubicin (daunomycin) and doxorubicin (adriamycin) are the best known. The general chemical structure of these compounds is shown in Fig. 11.9.7. These compounds are readily separated using reversed phase chromatography (Eksborg et al., 1978; Kono et al., 1979) with a buffered aqueous mobile phase containing an organic solvent modifier. An increase in the aqueous content of the mobile phase increases retention and optimal resolution is obtained at 25–30% water. Increasing the pH of the mobile phase increases the retention of compounds containing amino sugars (doxorubicin and daunorubicin) whereas it has little effect on the aglycones (adriamycinone and daunomycinone). This difference can be used to optimise the selectivity of the mobile phase.

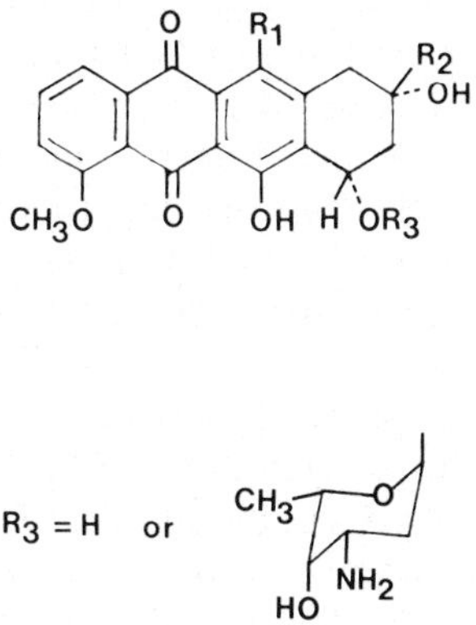

Fig. 11.9.7. Chemical representation of the general structure of anthracyclins.

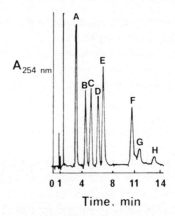

Fig. 11.9.8. Separation of an anthracyclin mixture. Chromatographic conditions: column, μBondapak C-18; mobile phase, 28% acetonitrile in aqueous potassium dihydrogen orthophosphate (8 g/l) acidified with citric acid to pH 3.0; flow rate, 5 ml/min; temperature, 60 °C; detection, UV at 254 nm. Peaks: A, doxorubicin; B, 13-dihydrodaunorubicin; C, adriamycinone; D, daunorubicin; E, 13-dihydrodaunomycinone; F, daunomycinone; G, 14-bromodaunorubicin; H, 14-bromodaunorubicin, 13-methylacetal. Reproduced from Alemanni et al. (1982), with permission.

Resolution of the anthracyclins is enhanced at elevated temperatures.

Accurate quantitation of anthracyclins in fermentation broths is important to facilitate determination of the ideal harvest time. Extraction of anthracyclins from these broths uses an initial acid treatment at pH 1.5 (Alemanni et al., 1982) and subsequent filtration or solvent extraction with acetonitrile–water (80 : 20). Samples can then be filtered for direct injection onto an HPLC column. This procedure has been used to examine ten anthracyclins produced in a fermentation broth. HPLC was carried out on a C_{18} reversed phase support in combination with an acetonitrile–aqueous phosphate buffer mobile phase (Fig. 11.9.8).

To study the pharmacokinetic properties of anthracyclins an HPLC technique to analyse the concentrations of the parent compound and its metabolites is required. Thus, the analysis of plasma samples of the anthracyclins utilised an initial extraction with an organic solvent such as chloroform in which the aqueous phase was maintained at pH 8.4. Back extraction into an acidic phosphate buffer was used to prevent interference from aglycone metabolites. HPLC analysis of the extracted anthracyclin and its metabolites was carried out on a μBondapak phenyl column eluted with acetonitrile–aqueous phosphate buffer (35:65) and was quantitated using fluorimetric detection (Moro et al., 1983).

11.9.7. Other antibiotics

The examples mentioned in this section are not intended to be exhaustive but merely to indicate to the reader other possibilities, whilst referring to more thorough texts.

The review by Nilsson-Ehle (1983) refers to a number of antibiotics not referred to previously and includes chloramphenicol, sulphonamides, trimethoprim, nitroimidazoles, rifampicin, 5-fluorocytosine, griseofulvin, acyclovir and antimalarial agents such as chloroquine. HPLC separations of other antibiotics have recently been described and include the macrolides (Bens et al., 1982a, 1982b; Veiga et al., 1983), norfloxacin (Pauliukonis et al., 1984), antibacterial agent AT2266 (Nakamura et al., 1983), methylparaben and derivatives (Radus and Gyr, 1983) and antimycin (Abidi, 1982).

Appendix

The following is a list of manufacturers and suppliers of HPLC and ancillary equipment. List I is restricted to those in the United Kingdom. Only a short list of the major suppliers in the US is presented in List 2; for more detailed information readers should refer to McNair (1984). The code numbers are used to refer to the different facilities available from each supplier:

1. Completely automated systems.
2. Stationary phases, columns.
3. Liquid delivery pumps.
4. Metering pumps.
5. Syringe driven pumps.
6. UV Detectors (*including diode arrray).
7. Fluorescence detectors.
8. Refractive index, infra-red detectors.
9. Radioactivity detectors.
10. Photoelectric, coulombic detectors.
11. Fraction collectors.
12. Ancillary equipment.

List I

Alltech Associates Applied Science Ltd.	New St., Carnforth, Lancashire LA5 9BX	2, 3, 12
Alpha-Bayer Ltd.	P.O. Box 4, Marlborough, Wiltshire	1, 2, 12
Anachem Ltd.	Charles St., Luton, Bedfordshire LU2 0EB	1, 2, 3, 4, 6, 7, 8, 11, 12
Analytical Instruments	London Rd., Pampisford, Cambridge CB2 4EF	1, 12

Analytical Supplies Ltd.	Little Eaton, Derby DE2 5DR	3, 4, 11, 12
Anderman and Company Ltd.	Kingston upon Thames, Surrey KT2 6NH	2, 12
Applied Biosystems	Birchwood Science Park, Warrington, Cheshire	1, 5, 12
Applied Chromatography Systems Ltd.	Concorde St., Luton, Bedfordshire LU2 0JE	1, 3, 4, 5, 6, 7, 8, 12
Austen (Charles) Pumps Ltd.	Royston Rd., Weybridge, Surrey KT14 7PB	3, 12
BDH Chemicals Ltd.	Broom Rd., Poole, Dorset BH12 4AN	1, 2, 12
Baird and Tatlock Ltd.	P.O. Box 1, Romford, Essex RM1 1HA	1, 3, 4, 6, 7, 11, 12
Beckman-RIIC Ltd.	Progress Rd., High Wycombe, Buckinghamshire HP12 4JL	1, 2, 3, 4, 6, 7, 8, 9, 12
Bio-Rad Labs. Ltd.	Caxton Way, Watford, Hertfordshire WD1 8RP	1, 2, 3, 6, 12
Boro Laboratories Ltd.	Aldermaston, Berkshire RG7 4QU	3, 12
Cecil Instruments Ltd.	Milton, Cambridge CB4 4AZ	1, 6, 7, 8, 12
Chemlab Instruments Ltd.	Upminster Rd., Hornchurch, Essex RM11 3XJ	2, 3, 8, 11, 12
Cherwell Laboratories Ltd.	Bicester, Oxon OX6 7XB	5, 12
Chrompack UK	Shrubbery Rd., London SW16	2, 12
Davidson and Hardy Ltd.	Antrim Rd., Belfast BT15 3BL	1, 2, 3, 4, 6, 7, 12
Dionex (UK) Ltd.	Eelmoor Rd., Farnborough, Hantshire GU14 7QN	1, 12
Du Pont (UK) Ltd.	Wedgwood Way, Stevenage, Hertfordshire SG1 4QN	1, 2, 6, 7, 8, 12
Dyson Instruments Ltd.	Hetton Lyons Industrial Estate, Hetton, Tyne and Wear DH5 0RN	1, 6, 7, 11, 12

EDT Research	14 Trading Estate Rd., London NW10 7LU	1, 8, 10, 12
Field Analytical Co. Ltd.	P.O. Box 113, Weybridge, Surrey KT13 8BJ	1, 2, 11, 12
Flygt Pumps Ltd.	Colwick, Nottingham NG4 2AN	3
Frost Instruments Ltd.	Fishponds Rd., Wokingham, Berkshire RG11 2QA	1, 6, 7, 12
Griffin and George	Ealing Rd., Alperton, Wembley, Middlesex HA0 1HJ	2, 3, 8, 9, 11, 12
HPLC Technology Ltd.	10 Waterloo St. West, Macclesfield, Cheshire SK11 6PJ	1, 2, 3, 6, 7, 12
Harrow Scientific Ltd.	Hagden Lane, Watford, Hertfordshire WD1 8LL	1, 12
Hewlett-Packard Ltd.	Nine Mile Ride, Wokingham, Berkshire RG11 3LL	1, 2, 6*, 7, 12
V.A. Howe & Co. Ltd.	St. Ann's Crescent, London SW18 2LS	2, 8, 9, 10, 12
Jones Chromatography Ltd.	Colliery Rd., Llanbradach, Mid-Glamorgan CF8 3QQ	2, 3, 4, 5, 6, 7, 8, 11, 12
Kemtronix Ltd.	High St., Compton, Berkshire RG16 0NL	1, 6, 7, 8, 12
Kontron Instruments Ltd.	Campfield Rd., St. Albans, Hertfordshire AL1 5JG	1, 3, 6, 7, 12
LDC/Milton Roy	Diamond Way, Stone Buisness Park, Stone, Staffordshire ST15 0HH	1, 2, 3, 4, 6, 7, 8, 12
LKB Instruments Ltd.	232 Addington Rd., Selsdon, Surrey CR2 8YD	1, 2, 3, 6*, 7, 8, 11, 12
Life Science Labs. Ltd.	Sedgewick Rd., Luton, Bedfordshire LU4 9DT	1, 3, 6, 7, 11, 12
Locarte Co. Ltd.	Wendell Rd., London W12	5, 7, 12
Mackay and Lynn Ltd.	2 West Bryson Rd., Edinburgh EH11 1EH	1, 3, 12

Magnus Scientific/ Magnus Data	68 Edison Rd., Rabans Lane, Aylesbury, Buckinghamshire HP19 3RS	1, 2, 4, 6, 7, 12
T.H. Mason & Sons	29 Parliament St., Dublin 2 Eire	1, 7, 11, 12
Meta Scientific	7 Fosters Grove, Windlesham, Surrey	1, 12
Millipore Ltd./ Waters Chromatography Division	11–15 Peterborough Rd., Harrow, Middlesex HA1 2YH	1, 2, 3, 6*, 7, 8, 10, 12
Northern Ireland Laboratory Services Ltd.	86 Botanic Avenue, Belfast BT7 1JR	1, 3, 6, 7, 8, 11, 12
Owens Polyscience Ltd.	34 Chester Rd., Macclesfield, Cheshire SK11 8DG	2, 3, 4, 8, 12
Perkin Elmer Ltd.	Post Office Lane, Beaconsfield, Buckinghamshire	1, 2, 6, 7, 8, 9, 10, 11, 12
Pharmacia Ltd.	Central Milton Keynes, Buckinghamshire MK9 3HP	1, 2, 3, 6, 11, 12
Phase Separations Ltd.	Deeside Industrial Estate, Queensferry, Clwyd CH5 2LR	2, 4, 12
Pierce (UK) Ltd.	36 Clifton Rd., Cambridge CB1 4ZR	2, 12
Polymer Laboratories Ltd.	Essex Rd., Church Stratton, Shropshire SY7 6AX	2, 3, 8, 12
Pye Unicam Ltd. Philips Analytical Division	York St., Cambridge CB1 2PX	1, 2, 3, 6, 7, 8, 9, 10, 11, 12
R.B. Radley & Co. Ltd.	London Rd., Sawbridge, Hertfordshire CM21 9JH	2, 5, 11, 12
Roth Scientific Co. Ltd.	Alexandra Rd., Farnborough, Hantshire RG21 3EY	1, 2, 6, 7, 8, 12
Scientific Supplies Co. Ltd.	Vine Hill, London EC1R 5EB	2, 3, 4, 11, 12
Spectra Physics Ltd.	17 Brick Noll Park, St. Albans, Hertfordshire AL1 5UF	1, 3, 6, 7, 12

Supelchem UK	London Rd., Sawbridge, Hertfordshire CM21 9JH	2, 11, 12
Technicol Ltd.	Brook St., Higher Hillgate, Stockport, Cheshire SK1 3HS	1, 2, 3, 6, 7, 8, 11, 12
Uniscience Ltd.	St. Ann's Crescent, Wandsworth, London SW18 2LS	2, 12
Varian Associates Ltd.	Manor Rd., Walton-on-Thames, Surrey	1, 2, 3, 6 *, 7, 8, 12
Watson-Marlow Ltd.	Falmouth, Cornwall TR11 4RU	3, 4, 12
Whatman Labsales Ltd.	Coldred Rd., Maidstone, Kent ME15 9XN	1, 2, 12

List 2

Beckman Altex Inc.	2350 Camino Ramon, San Ramon, CA 94121	1, 2, 3, 6, 7, 8, 9, 12
Bio-Rad Laboratories	1414 Harbour Way. Richmond, CA 94804	1, 2, 3, 6, 8, 12
Dionex Corp.	1228 Titan Way, Sunnyvale, CA 94086	1, 2, 3, 6, 7, 10, 12
EM Science Hitachi	111 Woodcrest Rd., Cherry Hill NJ 08034	1, 3, 6 *, 7, 12
Gilson Int.	P.O. Box 27, 3000 W. Beltline, 32 Middleton, WI 53562	1, 2, 3, 6, 8, 12
Hewlet-Packard Co. Analytical Products	3000 Hanover St., MS20B3 Palo Alto, CA 94304	1, 3, 6 *, 7, 8, 12
IBM Instruments Inc.	Orchard Park, P.O. Box 332 Danbury, CT 06810	1, 3, 6, 7, 8, 10, 12
Isco Inc.	P.O. Box 5347, 4700 Superior, Lincoln, NE 68505	1, 3, 6, 12
Kratos Analytical	170 Williams Dr., Ramsey, NJ 07446	1, 3, 6, 12

LCD/Milton Roy	P.O. Box 10235, Riviera Beach, FL 33404	1, 3, 6, 7, 8, 10, 12
LKB Instruments Inc.	9319 Gaither Rd., Gaithersburg, MD 20877	1, 2, 3, 6*, 7, 8, 12
Perkin-Elmer Corp.	761 Main Ave., Norwalk, CT 06859-0012	1, 3, 6, 7, 8, 12
Pharmacia Inc.	800 Centennial Ave., Piscataway, NJ 08854	1, 2, 6, 12
Shimadzu Scientific Inc.	7102 Riverwood Dr., Columbia, MA 21046	1, 3, 6*, 7, 8, 12
Spectra-Physics Autolab Division	3333 North First St., San Jose, CA 95134	1, 3, 6, 8, 12
Varian Instrument Group	220 Humboldt Court, Sunnyvale, CA 94089	1, 3, 6*, 7, 8, 10, 12
Waters Chromatography Division Millipore Corp.	34 Maple St., Milford, MA 01757	1, 2, 3, 6,*, 7, 8, 10, 12

References

Abbott, S.R. (1980) J. Chromatogr. Sci. *18*, 540–550.

Abbott, S.R., Berg, J.R., Achener, P. and Stevenson, R.L. (1973) J. Chromatogr. *126*, 241–249.

Abeydeera, W.P.P. (1983) Proc. Australian Soc. Sugar Cane Technologists, 171–186.

Abidi, S.L. (1982) J. Chromatogr. *234*, 187–200.

Abood, L.G., Salem, N., MacNeil, M. and Butler, M. (1978) Biochim. Biophys. Acta *530*, 35–38.

Agarwal, R.P., Major, P.P. and Kufe, D.W. (1982) J. Chromatogr. *231*, 418–424.

Aitzetmuller, K. (1978) J. Chromatogr. *156*, 354–358.

Aitzetmuller, K. (1980) Chromatographia *13*, 432–436.

Aitzetmuller, K. (1983) Prog. Lipid Res. *21*, 171–193.

Aitzetmuller, K. and Koch, J. (1978) J. Chromatogr. *145*, 195–202.

Alam, I., Smith, J.B., Silver, M.J. and Ahern, D. (1982) J. Chromatogr. *234*, 218–221.

Alemanni, A., Breme, U. and Vigevani, A. (1982) Proc. Biochem. May/June, 9–12.

Allen, B.A. and Nauman, R.A. (1975) J. Chromatogr. *190*, 241–247.

Allenmark, S. and Hedman, L. (1979) J. Liq. Chromatogr. *2*, 277–281.

Alpenfels, W.F., Mathews, R.A., Modden, D.C. and Newsom, A.E. (1982) J. Liq. Chromatogr. *5*, 1711–1715.

Ambler, M.R. and MacIntyre, D. (1975) J. Polymer Sci. *13*, 589–592.

Andersen, G.H. and Loevey, E.M.K. (1974) Prostaglandins *6*, 361–374.

Anderson, G.M. and Young, J.G. (1981) Life Sci. *28*, 507–517.

Anderson, J.E., Bond, A.M., Heritage, I.D., Jones, R.D. and Wallace, G.G. (1982) Anal. Chem. *54*, 1702–1705.

Anderson, F.S., Westcott, J.Y., Zirrolli, J.A. and Murphy, R.C. (1983) Anal. Chem. *55*, 1837–1839.

Ang, C.Y.W. and Moseley, J. (1980) J. Agric. Food Chem. *28*, 483–486.

Anhalt, J.P. (1977) Antimicrob. Agents Chemother. *11*, 651–656.

Anhalt, J.P. and Brown, S.D. (1978) Clin. Chem. *24*, 1940–1947.

Annesley, T., Grachero, D., Wilherson, K., Grekin, J. and Ellis, C. (1984) J. Chromatogr. *305*, 199–203.

Ansari, G.A.S. and Smith, L.L. (1979) J. Chromatogr. *175*, 307–315.

Armstrong, D.W. and Nome, F. (1981) Anal. Chem. *53*, 1662–1666.

Asmus, P.A. and Freed, C.R. (1979) J. Chromatogr. *169*, 303–311.

Asmus, P.A., Low, C. and Novotny, M. (1976) J. Chromatogr. *123*, 109–116.

Assenza, S.P. and Brown, P.R. (1983) J. Chromatogr. *282*, 477–486.

Assenza, S.P., Brown, P.R. and Goldberg, A.P. (1983) J. Chromatogr. *277*, 305–307.

Aszalos, A. and Aquilar, A. (1984) J. Chromatogr. *290*, 83–96.

Aszalos, A., Alexander, T. and Margosis, M. (1982a) Trends Anal. Chem. *1*, 387–393.

Aszalos, A., Haneke, C., Hayden, M. and Crawford, J. (1982b) Chromatographia *15*, 367–393.

Atwood, J.G. and Goldstein, J. (1980) J. Chromatogr. Sci. *18*, 650–654.

Au, J.L.-S., Weintjes, M.G., Luccioni, C.M. and Rustum, Y.M. (1982) J. Chromatogr. *228*, 245–256.

Azoulay, M., Desbene, P.L., Frappier, F. and Georges, Y. (1984) J. Chromatogr. *303*, 272–277.

Baenziger, J.U. and Natowicz, M. (1981) Anal. Biochem. *112*, 357–361.

Bailie, A.G., Wilson, T.D., O'Brien, R.K., Beebe, J.M., McCosh-Lilie, E.J. and Hill, D.W. (1982) J. Chromatogr. Sci. *20*, 466–470.

Barden, A.J. (1983) Anal. Biochem. *135*, 52–57.

Barton, F.E., Windham, W.R. and Himmelsbach, D.S. (1982) J. Agric. Food Chem. *30*, 1119–1123.

Batley, M., Redmond, J.W. and Tseng, A. (1982) J. Chromatogr. *253*, 124–128.

Batta, A.K., Dayal, V., Colman, R.W., Sinha, A.K., Shefa, S. and Salen, G. (1984) J. Chromatogr. *284*, 257–260.

Battaglia, R. and Frohlich, D. (1980) Chromatographia *13*, 428–431.

Baugh, C.M. and Krundieck, C.L. (1971) Ann. N. Y. Acad. Sci. *186*, 28–40.

Baust, J.G., Lee, R.E. and James, H. (1982) J. Liq. Chromatogr. *5*, 767–779.

Baust, J.G., Lee, R.E., Rojas, R.R., Hendrix, D.L., Friday, D. and James, H. (1983) J. Chromatogr. *261*, 65–75.

Bayer, E., Skutelsky, E. and Wilchek, M. (1979) Methods Enzymol. *62*D, 308–338.

Bayer, E., Albert, K., Nieder, M., Grom, E., Wolff, G. and Rindiisbacher, M. (1982) Anal. Chem. *54*, 1747–1750.

Bens, G.A., Crombez, E. and De Moerloose, P. (1982a) J. Liq. Chromatogr. *5*, 1449–1465.

Bens, G.A., Van den Bossche, W., Van der Weken, G. and De Moerloose, P. (1982b) J. Chromatogr. *248*, 312–317.

Berendsen, G.E. and De Galan, L. (1980) J. Chromatogr. *196*, 21–37.

Berendsen, G.E., Pikaart, K.A., De Galan, L. and Olieman, C. (1980) Anal. Chem. *52*, 1990–1993.

Bergh, M.L.E., Koppen, P.L., Van den Eijnden, Arnarp, J. and Lonngren, J. (1983) Carbohydr. Res. *117*, 275–278.

Bezard, J.A. and Ouedraogo, M.A. (1980) J. Chromatogr. *196*, 279–293.

Bidlingmeyer, B.A. and Warren, F.V. (1982) Anal. Chem. *54*, 2351–2356.

Bidlingmeyer, B.A., Deming, S.N., Price, W.P., Sachok, B. and Petrusek, M. (1979) J. Chromatogr. *186*, 419–424.

Bij, K.E., Horvath, C., Melander, W.R. and Nahum, A. (1981) J. Chromatogr. *203*, 65–84.

Binder, H. (1980) J. Chromatogr. *189*, 414–420.

Birmingham, B.K. and Green, D.S. (1983) J. Pharm. Sci. *72*, 1306–1309.

Blakely, C.R. and Vestal, M.L. (1983) Anal. Chem. *55*, 750–754.

Blank, L. (1976) J. Chromatogr. *117*, 35–40.

Blank, M.L. and Snyder, F. (1983) J. Chromatogr. *273*, 415–420.

Blaschek, W. (1983) J. Chromatogr. *256*, 157–163.

Blom, C.P., Deierkauf, F.A. and Riemersma, J.C. (1979) J. Chromatogr. *171*, 331–338.

Boersma, A., Lamblin, G., Degand, P. and Roussel, P. (1981) Carbohydr. Res. *94*, C7–C9.

Boeynaems, J.M., Brash, A.R., Oates, J.A. and Hubbard, W.C. (1980) Anal. Biochem. *104*, 259–267.

Borch, R.F. (1975) Anal. Chem. *47*, 2437–2439.

Bowman, S.A. (1982) Anal. Chem. *54*, 328A–332A.

Brando, S.C.C., Richmond, M.L., Gray, J.I., Morton, I.D. and Stine, C.M. (1980) J. Food Sci. *45*, 1492–1493.

Bratin, K. and Kissinger, P.T. (1981) J. Liq. Chromatogr. *4* (Suppl. 2), 321–357.

Bremer, E.G., Gross, S.K. and McLuer, R.H. (1979) J. Lipid Res. *20*, 1028–1035.

Breter, H.-J., Seibert, G. and Zahn, R.K. (1977) J. Chromatogr. *140*, 251–256.

Brewer, S.J., Dickerson, C.D., Ewbank, J.J. and Fallon, A. (1986) J. Chromatogr. 362, 443–449.

Bristol, D.W. (1980) J. Chromatogr. *188*, 193–204.

Bristow, P.A. (1976) Liquid Chromatography in Practice. Hetp, Cheshire.

Brons, C. and Olieman, C. (1983) J. Chromatogr. *259*, 79–86.

Brown, E.G., Newton, R.P. and Shaw, N.M. (1982) Anal. Biochem. *123*, 378–388.

Brugman, W.J.Th., Heemstra, S. and Kraak, J.C. (1982) Chromatographia *15*, 282–288.

Brust, O.-E., Sebastian, I. and Halasz, I. (1973) J. Chromatogr. *83*, 15–24.

Buchberger, W., Winsauer, K. and Breitwieser, Ch. (1983) Fresenius Z. Anal. Chem. *315*, 518–520.

Buck, M., Connick, M. and Ames, B.N. (1983) Anal. Biochem. *129*, 1–13.

Buckee, G.K. and Long, D.E. (1982) J. Am. Soc. Brewing Chemists *40*, 137–140.

Buhs, R.P., Maxim, T.E., Allen, N., Jacob, T.A. and Wolf, F.J. (1974) J. Chromatogr. *99*, 609–618.

Bui-Nguyen, M.H. (1980) J. Chromatogr. *196*, 163–165.

Bussell, N.E., Gross, A. and Miller, R.A. (1979) J. Liq. Chromatogr. *2*, 1337–1342.

Buthriss, J.L. and Diplock, A.J. (1984) Methods Enzymol. *105*, 131–138.

Cassini, A., Martini, F., Nieri, S., Ramarli, D., Franconi, F. and Surrenti, C. (1982) J. Chromatogr. *249*, 187–192.

Caude, M. and Foucault, A. (1979) Anal. Chem. *51*, 459–462.

Caude, M.H., Jardy, A.P. and Rosset, R.H. (1984) CRC Handbook of HPLC Separation of Amino Acids, Peptides and Proteins. *1*, 411–422.

Chang, C.A. and Tu, C.-F. (1982) Anal. Chem. *54*, 1179–1182.

Chang, S.H., Gooding, K.M. and Regnier, F.E. (1976) J. Chromatogr. *120*, 321–323.

Chang, J.Y., Brauer, D., Wittmann, J. and Liebold, B. (1980) FEBS Lett. *93*, 205–214.

Chang, J.Y., Knecht, R. and Braun, D.G. (1984) Methods in Protein Sequence Analysis. Elzing, M. (Ed.). Humana Press, Clifton, N.J. pp. 113–119.

Cheetham, N.W.H., Sirimanne, P. and Day. W.R. (1981) J. Chromatogr. *207*, 439–444.

Chen, S.S.-H. and Kou, A.Y. (1982) J. Chromatogr. *227*, 25–31.

Chen, C.-C. and McGinnis, G.D. (1983) Carbohydr. Res. *122*, 322–326.

Chen, S.S.-H., Kou, A.Y. and Chen, H.-H.Y. (1981) J. Chromatogr. *208*, 339–346.

Chmielowiec, J. (1981) J. Chromatogr. Sci. *19*, 296–307.

Chmielowiec, J. and George, A.E. (1980) Anal. Chem. *52*, 1154–1157.

Chow, F.K. and Grushka, E. (1977) Anal. Chem. *49*, 1756–1759.

Chow, F.K. and Grushka, E. (1979) J. Chromatogr. *185*, 361–373.

Christie, W.W. (1982) In: Lipid Analysis, 2nd. edn., Pergammon Press, Oxford.
Clarke, M.J., Coffey, K.F., Perpall, H.J. and Lyon, J. (1982) Anal. Biochem. *122*, 404–411.
Cohn, W.E. (1949) Science *109*, 377–378.
Colin, H. and Guiochon, G. (1977) J. Chromatogr. *141*, 289–312.
Colin, H., Guichon, G., and Siouffi, A. (1979) Anal. Chem. *51*, 1661–1666.
Colin, H., Guiochon, G. and Jandera, P. (1983a) Anal. Chem. *55*, 442–446.
Colin, H., Krstulovic, A., Guiochon, G. and Yun, Z. (1983b) J. Chromatogr. *255*, 295–309.
Colonna, A., Russo, T., Esposito, F., Salvatore, F. and Cimino, F. (1983) Anal. Biochem. *130*, 19–26.
Compton, B.J. and Purdy, W.C. (1982) Acta Chim. Acta *141*, 405–410.
Consden, R., Gordon, A.H. and Martin, A.J.P. (1944) Biochem. J. *38*, 224–232.
Cooke, N.H., Archer, B.G., Olsen, K. and Berick, A. (1982) Anal. Chem. *54*, 2277–2283.
Cooper, M.J. and Anders, M.W. (1974) Anal. Chem. *46*, 1849–1852.
Cooper, A.R., Hughes, A.J. and Johnson, J.F. (1975) J. Appl. Polymer Sci. *19*, 435–437.
Crawford, C.G., Plattner, R.D., Sessa, D.J. and Rackis, J.J. (1980) Lipids *15*, 91–94.
CRC Handbook of Chemistry and Physics. CRC Press, Florida.
Crombeen, J.P., Kraak, J.C. and Poppe, H. (1978) J. Chromatogr. *67*, 219–230.
Crossley, K.R., Rotschafer, J.C., Chern, M.M., Mead, K.E. and Zaske, D.E. (1980) Antimicrob. Agents Chemother. *17*, 654–657.
Crowell, E.P. and Burnett, B.B. (1967) Anal. Chem. *39*, 121–124.
Crowther, J.B., Jones, R. and Hartwick, R.A. (1981) J. Chromatogr. *217*, 479–490.
Crowther, J.B., Caronia, J.P. and Hartwick, R.A. (1982) Anal. Biochem. *124*, 65–73.
Cuatrecasas, P.M., Wilchek, M. and Anfinsen, C.B. (1968) Proc. Natl. Acad. Sci. U.S.A. *61*, 636–639.
Curstedt, T. and Sjovall, J. (1974) Biochim. Biophys. Acta *360*, 24–37.
D'Amboise, M., Noel, D. and Hanai, T. (1980) Carbohydr. Res. *79*, 1–10.
Darney, K.J., Wing, T.-Y. and Ewing, L.L. (1983) J. Chromatogr. *257*, 81–90.
Darwish, A.A. and Prichard, R.K. (1981) J. Liq. Chromatogr. *4*, 1511–1524.
Davankov, V.A. and Semechkin, A.V. (1977) J. Chromatogr. *141*, 313–319.
Davankov, V.A. and Zolotarev, Y.A. (1978) J. Chromatogr. *155*, 285–311.
Davankov, V.A., Kurganov, A.A. and Bochkov, A.S. (1983) Advances in Chromatography (Giddings, J.C., Gusshka, E., Cayes, J. and Brown, H., eds.) Marcel Dekker, New York, NY.
Davis, G.E., Suits, R.D., Kuo, K.C., Gehrke, C.W., Waalkes, T.P. and Borek, A. (1977) Clin. Chem. *23*, 1427–1435.
Davis, T.P., Gehrke, C.W., Cunningham, T.D., Kuo, K.C., Gerhardt, K.O., Johnson, H.D. and Williams, C.H. (1978) Clin. Chem. *24*, 1317–1324.
Davis, G.E., Gehrke, C.W., Kuo, K.C. and Agris, P.F. (1979) J. Chromatogr. *173*, 281–298.
Davis, T.P., Shoemaker, H., Chen, A. and Yamamura, H.I. (1982) Life Sci. *30*, 971–987.
DeAbreu, R.A., Van Baal, J.M., De Bruyn, C.H.M.M., Bakkeren, J.A.J.M. and Schretlen, E.D.A.M. (1982) J. Chromatogr. *229*, 67–75.
Dean, P.D.G., Johnson, W.S. and Middle, F.A. (1985) In: Affinity Chromatography, A. Practical Approach (Rickwood, D. and Hames, B.D., eds). IRL Press. Lancaster.
Debetto, P. and Bianchi, V. (1983) J. High Res. Chromatogr. Commun. *6*, 117–122.
Deelder, R.S., Linssen, H.A.J., Konijnendijk, J. and Van de Venne, J.L.M. (1979) J. Chromatogr. *185*, 241–257.

Delia, T.J., Kirt, D.D. and Drach, J.C. (1982) J. Chromatogr. *243*, 173–177.
De Ligny, C.L., Spanjer, M.C., Van Howelingen, J.C. and Weesie, H.M. (1984) J. Chromatogr. 301, 311–324.
Delort, A.M., Derbyshire, R., Duplaa, A.M., Guy, A., Molko, D. and Teoule, R. (1984) J. Chromatogr. *283*, 462–467.
DeLuca, P.P. and Papadimitriou, D. (1983) J. Parent. Sci. Technol. *37*, 125–128.
Desai, I.D. (1984) Methods Enzymol. *105*, 138–147.
Desbene, P.L., Coustal, S. and Frappier, F. (1983) Anal. Biochem. *128*, 359–362.
De Stefano, J.J. and Beachell, H.C. (1972) J. Chromatogr. Sci. *10*, 654–656.
De Stefano, J.J. and Kirkland, J.J. (1975) Anal. Chem. *47*, 1103A.
Dewaele, C. and Verzele, M. (1983) J. Chromatogr. *260*, 13–21.
Diala, E.S. and Hoffman, R.M. (1982) Biochem. Biophys. Res. Commun. *107*, 19–26.
Dieter, D.S. and Walton, H.F. (1983) Anal. Chem. *55*, 2109–2112.
Distler, W. (1980) J. Chromatogr. *192*, 240–246.
Dizdaroglu, M. (1985) J. Chromatogr. *334*, 49–69.
Do, U.H., Pei, P.T. and Minard, R.D. (1981) Lipids *16*, 855–862.
Dolphin, R.J. and Pergande, P.J. (1977) J. Chromatogr. *143*, 267–274.
Dua, V.K. and Bush, C.A. (1983) Anal. Biochem. *133*, 1–8.
Dufek, P. and Smolkova, E. (1983) J. Chromatogr. 257, 247–254.
Duncan, I.W., Culbreth, P.H. and Burtis, C.A. (1979) J. Chromatogr. *162*, 281–292.
Dunham, E.W. and Anders, M.W. (1973) Prostaglandins, *4*, 85–92.
Durst, H.D., Milano, H., Kikta, E.J., Connelly, S.A. and Grushka, E. (1975) Anal. Chem. *47*, 1797–1799.
Dutch, D.S., Bowers, S.W. and Nichol, C.A. (1983) Anal. Biochem. *130*, 385–390.
Ebling, W.F. Szefler, S.J. and Jusko, W.J. (1984) J. Chromatogr. *305*, 271–280.
Eckers, C., Cuddy, K.K. and Henion, J.D. (1983) J. Liq. Chromatogr. 6, 2383–2409.
Edy, V.B., Billiau, A. and De Sommer, P. (1977) J. Biol. Chem. *252*, 5934–5936.
Egberg, D.C., Heroff, J.C. and Potter, R.H. (1977) J. Agric. Food Chem. *25*, 1127–1131.
Eisenberg, F. Jr. (1971) Carbohydr. Res. *19*, 135–138.
Eksborg, S., Ehrsson, H., Andersson, B. and Beran, M. (1978) J. Chromatogr. *153*, 211–218.
Ekstrom, B. and Jackobson G. (1984) Anal. Biochem. *142*, 134–139.
El-Hamdy, A.H. and Perkins, E. (1981) J. Am. Oil Chem. Soc. *58*, 867–872.
Eling, T., Tainer, B., Ally, A. and Warnock, R. (1982) Methods Enzymol. *86*, 511–517.
Ellis, G.P. and Honeyman, J. (1955) Adv. Carbohydr. Chem. *10*, 95–168.
Endo, M., Suzuki, K., Schmid, K., Fownet, B., Karamanos, Y., Montreuil, J., Dorland, L., Van Halbeek, H. and Vliegenthart, J.F.G. (1982) J. Biol. Chem. *257*, 8755–8760.
Engel, L.L. (ed.) (1963) The Physical Properties of Steroid Hormones. Pergamon Press, Oxford.
Engelhardt, H. and Ahr, G. (1981) Chromatographia *14*, 227–233.
Engstrom, R.C. (1982) Anal. Chem. *54*, 2310–2314.
Ericson, A., Niklasson, F. and De Verdier, C.-H. (1983) Clin. Chim. Acta *127*, 47–59.
Erlich, M. and Erlich, K. (1979) J. Chromatogr. Sci. *17*, 531–535.
Essers, L. (1984) J. Chromatogr. *305*, 345–352.
Ettre, L. (1980) High Perf. Liq. Chromatogr. *1*, 1–74.
Ettre, L.S. and Horvath, C. (1975) Anal. Chem. *47*, 422A–444A.
Evans, J.E. and McCluer, R.H. (1972) Biochim. Biophys. Acta *270*, 565–569.
Evans, M.B., Dale, A.D. and Little, C.J. (1980) Chromatographia *13*, 5–10.
Felice, L.J., Felice, J.D. and Kissinger, P.T. (1978) J. Neurochem. *31*, 1461–1465.

Fenn. R.J., Siggia, S. and Curran, D.J. (1978) Anal. Chem. *50*, 1067–1072.
Fishman, S. (1975) J. Pharm. Sci. *64*, 674–680.
Fitzpatrick, F.A. (1976) Anal. Chem. *48*, 499–502.
Fitzpatrick, F.A. and Siggia, S. (1973) Anal. Chem. *45*, 2310–2314.
Fitzpatrick, F.A., Siggia, S. and Dingman, J. (1972) Anal. Chem. *44*, 2211–2216.
Fitzpatrick, F.A., Wynalda, M.A. and Kaiser, D.G. (1977) Anal. Chem. *49*, 1032–1035.
Fleet, B. and Little, C.J. (1974) J. Chromatogr. Sci. *12*, 747–757.
Floridi, A., Palmerini, C.A. and Fini, C. (1977) J. Chromatogr. *138*, 203–212.
Foltz, A.K., Yeransian, J.A. and Sloman, K.G. (1983) Anal. Chem. *55*, 164R–196R.
Frank, H.S. and Evans, M.W. (1945) J. Chem. Phys. *13*, 507–509.
Freukel, E., Prough, R. and Kitchens, R.C. (1980) Methods Enzymol. *67*, 31–40.
Frey, B.M. and Frey, F.J. (1982) Clin. Chem. *28*, 689–692.
Friedlander, J., Fischer, D.G. and Rubinstein, M. (1984) Anal. Biochem. *137*, 115–119.
Fritz, H.-J., Belagaje, R., Brown, E.L., Fritz, R.H., Jones, R.A., Lees, R.G. and
    Khorana, H.G. (1978) Biochemistry *17*, 1257–1267.
Frolik, C.A., Dart, L.L. and Sporn, M.B. (1982) Anal. Biochem. *125*, 203–209.
Fullmer, C.S. (1984) Anal. Biochem. *142*, 336–339.
Furakawa, H., Mori, Y., Takeuchi, Y. and Ito, K. (1977) J. Chromatogr. *136*, 428–431.
Gait, M.J., Matthes, H.W.D., Singh, M., Sproat, B.S. and Titmas, R.C. (1982) Nucleic
    Acids Res. *10*, 6234–6254.
Gandelman, M.S. and Birks, J.W. (1983) Anal. Chim. Acta *155*, 159–171.
Gant, J.R., Dolan, J.W. and Snyder, L.R. (1979) J. Chromatogr. *185*, 153–177.
Garrett, C. and Santi, D.V. (1979) Anal. Biochem. *99*, 268–273.
Gehrke, C.W., Kuo, K.C. and Zumwalt, R.W. (1980) J. Chromatogr. *188*, 129–147.
Gehrke, C.W., Kuo, K.C., McCune, R.A. and Gerhardt, K.O. (1982) J. Chromatogr.
    230, 297–308.
Geiss, F. and Schlitt, H. (1976) J. Chromatogr. *82*, 5–6.
Geurts van Kessel, W.S.M., Hax, W.M.A., Demel, R.A. and De Gier, J. (1977) Biochim.
    Biophys. Acta *486*, 524–529.
Gfeller, J.C., Frey, G. and Frei, R.W. (1977) J. Chromatogr. *142*, 271–281.
Gil-Av, E. and Weinstein, S. (1984) CRC Handbook for the Separation of Amino
    Acids, Peptides and Proteins. *1*, 429–442.
Gilpin, R.K., Pachla, L.A. and Ranweller, J.S. (1983) Anal. Chem. *55*, 70R–87R.
Glad, M., Ohlsen, S., Hansson, L., Mansson, M.-O. and Mosbach, K. (1980) J.
    Chromatogr. *200*, 254–260.
Gnanasambandan, T. and Freiser, H. (1982) Anal. Chem. *54*, 2379–2380.
Goeddel, D.V., Kleid, D.G., Bolivar, F., Heyneker, H.L., Yansura, D.G., Crea, R.
    Hirose, T., Kraszewski, A., Itakura, K. and Riggs, A.D. (1979) Proc. Natl. Acad. Sci.
    U.S.A. *76*, 106–110.
Goehl, T.J., Sundaresan, G.M. and Vadlamenl, P.K. (1979) J. Pharm. Sci. *40*, 1374–1376.
Goetzl, E.J. and Sun, F.F. (1979) J. Exp. Med. *150*, 406–411.
Goldberg, A.P. (1982) Anal. Chem. *54*, 342–345.
Gotelli, G.R., Kabra, P.M. and Marton, L.J. (1977) Clin. Chem. *23*, 165–168.
Gotelli, G.R., Wall, J.H., Kabra, P.M. and Marton, L.J. (1981) Clin. Chem. *27*,
    441–443.
Goto, M., Sakurai, E. and Ishii, D. (1982) J. Chromatogr. *238*, 357–366.
Goto, M., Zou, G. and Ishii, D. (1983) J. Chromatogr. *275*, 271–281.
Graeve, L., Goemann, W., Foldi, P. and Kruppa, J. (1982) Biochem. Biophys. Res.
    Commun. *107*, 1559–1565.

Granneman, G.R. and Sennello, L.T. (1982) J. Chromatogr. *229*, 149–157.

Gray, W.R. and Hartley, B.S. (1963) Biochem. J. *89*, 379–380.

Green, K. and Samuelsson, B. (1964) J. Lipid Res. *5*, 117–121.

Greenspan, M.D. and Schroeder, E.A. (1982) Anal. Biochem. *127*, 441–448.

Grimble, G.K., Barker, H.M. and Taylor, R.H. (1983) Anal. Biochem. *128*, 422–428.

Grizzle, P.L. and Thomson, J.S. (1982) Anal. Chem. *54*, 1071–1078.

Grell, E., Lewitzki, E., Dehal, S., Oberbaumer, I., Raschdorf, F. and Richter, W.T. (1984) J. Chromatogr. *290*, 57–63.

Gross, R.W. and Sobel, B.E. (1980) J. Chromatogr. *197*, 79–85.

Haddad, P.R., Keating, R.W. and Low, G.K.C. (1982) J. Liq. Chromatogr. *5*, 853–867.

Hageman, J.H. and Kuehn, G.D. (1977) Anal. Biochem. *80*, 547–554.

Halasz, I. (1980) Anal. Chem. *52*, 1393A–1403A.

Halicioglu, T. and Sinanoglu, O. (1969) Ann. N.Y. Acad. Sci. *158*, 308–320.

Hall, R.H. (1971) In: The Modified Nucleosides in Nucleic Acids. Columbia University Press, New York, NY, pp. 281–294.

Halvorsen, O. and Skrede, S. (1980) Anal. Biochem. *107*, 103–108.

Hamberg, M., Svensson, J. and Samuelsson, B. (1975) Proc. Natl. Acad. Sci. U.S.A. *72*, 2994–2997.

Hamilton, P.B. (1963) Anal. Chem. *35*, 2055–2056.

Hamilton, J.G. and Karol, R.J. (1982) Prog. Lipid Res. *21*, 155–170.

Hammond, E.W. (1981) J. Chromatogr. *203*, 397–403.

Hansbury, E. and Scallen, T.J. (1982) Chromatogr. Sci. *16*, 253–276.

Hanson, V.L., Park, J.Y., Osborn, T.W. and Kiral, R.M. (1981) J. Chromatogr. *205*, 393–400.

Hara, I., Shiraishi, K. and Okazaki, M. (1982) J. Chromatogr. *239*, 549–557.

Hare, P.E. and Gil-Av, E. (1979) Science *204*, 1226–1228.

Harmenberg, J. (1983) J. Liq. Chromatogr. *6*, 655–666.

Hartwick, R.A. and Brown, P.R. (1975) J. Chromatogr. *112*, 651–657.

Hartwick, R.A., Grill, C.M. and Brown, P.R. (1979a) Anal. Chem. *51*, 34–37.

Hartwick, R.A., Krstulovic, A.M. and Brown, P.R. (1979b) J. Chromatogr. *186*, 659–676.

Hartwick, R.A., Assenza, S.P. and Brown, P.R. (1979c) J. Chromatogr. *186*, 647–658.

Hastings, C.R., Aue, W.E. and Larsen, F.N. (1971) J. Chromatogr. *60*, 329–344.

Haupt, W. and Pingoud, A. (1983) J. Chromatogr. *260*, 419–427.

Hearn, M.T.W. (1980) J. Liq. Chromatrogr. *3*, 1255–1276.

Hefti, F. (1979) Life Sci. *25*, 775–782.

Heftman, E. (1983) J. Chromatogr., Libr. *22B*, 191–222.

Heftman, E. and Lin, J.-T. (1982) J. Liq. Chromatogr. *5* (Suppl. I), 121–173.

Heinrikson, R.L. and Meredith, S.C. (1984) Anal. Biochem. *136*, 65–74.

Hemdan, S.E. and Porath, J. (1985a) J. Chromatogr. *323*, 247–254.

Hemdan, S.E. and Porath, J. (1985b) J. Chromatogr. *323*, 255–264.

Hemdan, S.E. and Porath, J. (1985c) J. Chromatogr. *323*, 265–272.

Henderson, R.J. and Griffin, C.A. (1981) J. Chromatogr. *226*, 202–207.

Hendrix, D.L., Lee, R.E., Baust, J.G. and James, H. (1981) J. Chromatogr. *210*, 45–53.

Hengen, N., Seiberth, V. and Hengen, M. (1978) Clin. Chem. *24*, 1740–1743.

Hennion, M.C., Picard, C., Cormbellas, C., Caude, M. and Rosset, R. (1981) J. Chromatogr. *210*, 211–228.

Henschen, A, Hupe, K.P., Lottspeich, F. and Voelter, W. (eds.) (1985) HPLC in Biochemistry, VCH, Weinheim, F.R.G.

Hermansson, J. (1978) J. Chromatogr. *152*, 437–442.
Herslof, B.G., Podlaha, O. and Toregard, B. (1979) J. Am. Oil Chem. Soc. *56*, 864–866.
Hesse, C. and Hovermann, W. (1973) Chromatographia *6*, 345–348.
Hodge, J.L. and Rossomando, E.F. (1980) Anal. Biochem. *102*, 59–62.
Hoffman, N.E. and Liao, J.C. (1977) Anal. Chem. *49*, 2231–2235.
Hogan, D.L., Kraemer, K.L. and Isenberg, J.I. (1983) Anal. Biochem. *127*, 17–24.
Holcombe, I.J. and Fusari, S.A. (1981) Anal. Chem. *53*, 607–609.
Honda, S., Takahashi, M., Araki, Y. and Kakehi, K. (1983) J. Chromatogr. *274*, 45–52.
Honma, Y. (1982) Ind. Health. *20*, 247–258.
Horne, D.W, Briggs, W.T. and Wagner, C. (1981) Anal. Biochem. *116*, 393–398.
Horvath, C. (ed.) (1983) High Performance Liquid Chromatography: Advances and Perspectives, Vol. III, Academic Press, New York, NY.
Horvath, C. and Lin, H. (1978) J. Chromatogr. *149*, 43–50.
Horvath, C. and Lipsky, S.R. (1966) Nature *211*, 748–750.
Horvath, C. and Melander, W. (1977) J. Chromatogr. Sci. *15*, 393–404.
Horvath, C. and Melander, W. (1978) Am. Lab. *10*, 17–36.
Horvath, C., Preiss, B.A. and Lipsky, S.R. (1967) Anal. Chem. *39*, 1422–1428.
Horvath, C., Melander, W. and Molnar, I. (1976) J. Chromatogr. *125*, 129–156.
Horvath, C., Melander, W. and Molnar, I. (1977a) Anal. Chem. *49*, 142–154.
Horvath, C. Melander, W., Molnar, I. and Molnar, P. (1977b) Anal. Chem. *49*, 2295–2298.
Howard, G.A. and Martin, A.J.P. (950) Biochem. J. *46*, 532–538.
Howard, S. and McCormick, D.B. (1981) J. Chromatogr. *208*, 129–134.
Hudson, T.J. and Allen, R.J. (1984) J. Pharm. Sci. *73*, 1–10.
Hull-Ryde, E.A., Cummings, R.G. and Lowe, J.E. (1983) J. Chromatogr. *275*, 411–417.
Hunkapiller, M.W. and Hood, L.E. (1978) Biochemistry *17*, 2124–2128.
Hunter, I.R., Walden, M.K. and Heftman, E. (1978) J. Chromatogr. *153*, 57–61.
Hunter, I.R., Walden, M.K. and Heftman, E. (1979) J. Chromatogr. *176*, 485–487.
Hupe, K.P. and Lauer, H.H. (1981) J. Chromatogr. *203*, 41–52.
Hurtubise, R.J., Hussain, A. and Silver, H.F. (1981) Anal. Chem. *53*, 1993–1997.
Ike, Y., Ikuta, S., Sato, M., Huang, T. and Itakura, K. (1983) Nucleic Acids Res. *11*, 477–488.
Ikeda, M., Shimada, K. and Sakaguchi, T. (1983) J. Chromatogr. *272*, 251–259.
Ikensya, S., Hiroshima, O., Ohmae, M. and Kawabe, K. (1980) Chem.-Pharm. Bull. *28*, 2941–2947.
Ikumo, Y., Maoka, T., Shumizu, M., Komori, T. and Matsumo, T. (1985) J. Chromatogr. *328*, 387–391.
Imai, K. (1975) J. Chromatogr. *105*, 135–141.
Imai, K. and Zamura, Z. (1978) Clin. Chim. Acta *85*, 1–7.
Ingerbretsen, O.C., Normann, D.T. and Flatmark, T. (1979) Anal. Biochem. *96*, 181–188.
Ishii, D. and Takeuchi, T. (1983) J. Chromatogr. *255*, 349–358.
Ishii, K., Sarai, K., Sanemori, H. and Kawasaki, T. (1979) Anal. Biochem. *97*, 191–195.
Ishii, D., Hibi, K., Asai, K., Nagaya, M., Mochizuki, K. and Mochida, Y. (1978) J. Chromatogr. *156*, 173–180.
Iwamori, M. and Moser, H.W. (1975) Clin. Chem. *21*, 725–729.
Iwamori, M. and Nagai, Y. (1978) Biochim. Biophys. Acta *528*, 257–259.
Iwamori, M., Costello, C. and Moser, H.W. (1979) J. Lipid Res. *20*, 86–96.
Jackman, G.P., Carson, V.J., Bobik, A. and Skews, H. (1980) J. Chromatogr. *182*, 277–284.

Jacobsen, D.W., Green, R., Quadros, E.V. and Montejano, Y.D. (1982) Anal. Biochem. *120*, 394–403.
Jandera, P., Colin, H. and Guiochon, G. (1982) Anal. Chem. *54*, 435–441.
Jensen, G.W. (1981) J. Chromatogr. *204*, 407–411.
Johansson, I.M., Wahlund, K.G. and Schill, G. (1978) J. Chromatogr. *149*, 281–296.
Jones, B.N. and Gilligan, J.P. (1983) Am. Biotechnol. Lab. *December*, 52–57.
Jones, B.N., Paabo, S. and Stein, S. (1981) J. Liq. Chromatogr. *4*, 565–586.
Jonvel, P., Andermann, G. and Bartholemy, J.F. (1983) J. Chromatogr. *281*, 371–376.
Jorgenson, J.W. and Guthrie, E.J. (1983) J. Chromatogr. *255*, 335–348.
Josic, D.J., Baumann, H. and Reutter, W. (1984) Anal. Biochem. *142*, 473–479.
Jost, W., Unger, K. and Schill, G. (1982) Anal. Biochem. *119*, 214–223.
Jungalwala, F.B., Turel, R.J., Evans, J.E. and McCluer, R.H. (1975) Biochem. J. *145*, 517–526.
Jungalwala, F.B., Evans, J.E. and McCluer, R.H. (1976) Biochem. J. *155*, 55–60.
Jungalwala, F.B., Hayes, L. and McCluer, R.H. (1977) J. Lipid Res. *18*, 285–292.
Jungalwala, F.B., Evans, J.E. and McCluer, R.H. (1983) J. Lipid Res. *24*, 1380–1388.
Kadowaki, H., Bremer, E.G., Evans, J.E., Jungalwala, F.B. and McCluer, R.H. (1983) J. Lipid Res. *24*, 1389–1397.
Kaduce, T.L., Norton, K.C. and Spector, A.A. (1983) J. Lipid Res. *24*, 1398–1403.
Kaminski, M., Klawiter, J. and Kowalczyk, J.S. (1982) J. Chromatogr. *243*, 225–244.
Kannagi, R., Fukuda, M.N. and Hakomori, S. (1982) J. Biol. Chem. *257*, 4438–4442.
Karch, K., Sebastian, I. and Halasz, I. (1976) J. Chromatogr. *122*, 3–9.
Karger, B.L., Gant, J.R., Hartknopf, A. and Weiner, P.H. (1976) J. Chromatogr. *128*, 65–71.
Karger, B.L., Le Page, J.N. and Tanaka, N. (1980) High Perf. Liq. Chromatogr. *1*, 113–206.
Karlesky, D., Shelly, D.C. and Warner, I. (1981) Anal. Chem. *53*, 2146–2147.
Kasumoto, K.J. (1975) J. Nutr. Sci. Vitaminol. *21*, 117–120.
Kato, Y., Komiya, K., Sasaki, H. and Hashimoto, T. (1980) J. Chromatogr. *193*, 458–464.
Kato, Y., Komiya, K., and Hashimoto, T. (1982) J. Chromatogr. *246*, 13–18.
Kato, Y., Nakamura, K. and Hashimoto, T. (1983a) J. Chromatogr. *266*, 385–390.
Kato, Y., Sasaki, M., Hashimoto, T., Murotsu, T., Fukushige, S. and Matsubara, K. (1983b) J. Chromatogr. *265*, 342–346.
Kato, Y., Nakamura, K. and Hashimoto, T. (1984) J. Chromatogr. *294*, 207–212.
Kautsky, M.P. (ed.) (1982) Steroid Analysis by HPLC: Recent Applications. Chromatography Science Vol. 16. Marcel Dekker, New York NY.
Kawalek, J.C., Hwang, K.K. and Kelsey, M.I. (1981) Chromatographia *14*, 633–637.
Kawamoto, T. and Okada, E. (1983) J. Chromatogr. *258*, 284–288.
Kawasaki, T., Maeda, M. and Tsuji, A. (1979) J. Chromatogr. *163*, 143–150.
Kawasaki, T., Maeda, M. and Tsuji, A. (1982) J. Chromatogr. *233*, 61–67.
Keller, R., Oke, A., Mefford, I. and Adams, R.N. (1976) Life Sci. *19*, 995–1004.
Kessler, M.J. (1982) J. Liq. Chromatogr. *5*, 111–123.
Kier, L.B., Hall, L.H., Murray, N.J. and Randie, M. (1975) J. Pharm. Sci. *64*, 1971–1976.
Kimura, M., Fujita, T. and Itokawa, Y. (1982) Clin. Chem. *28*, 29–31.
Kirchner, J.G. (1973) J. Chromatogr. Sci. *11*, 180–183.
Kissinger, P.T. (1983) Chromatogr. Sci. *23*, 125–164.
Kissinger, P.T., Refshauge, C.J., Dreiling, R. and Adams, R.N. (1973) Anal. Lett. *6*, 465–477.

Kiuchi, K., Ohta, T. and Ebine, H. (1977) J. Chromatogr, *133*, 226–230.

Knapp, D.R. and Krueger, S. (1975) Anal. Lett. *8*, 603–606.

Kniep, B., Hunig, T.G., Fitch, F.W., Heuer, J., Kolsch, E. and Muhlradt, P.F. (1983) Biochemistry *22*, 251–255.

Knighton, D.R., Harding, D.R.K., Napier, J.R. and Hancock, W.S. (1982) J. Chromatogr. *249*, 193–198.

Knox, J.H. and Gilbert, M.T. (1979) J. Chromatogr. *186*, 405–409.

Knox, J.H. and Jurand, J. (1976) J. Chromatogr. *125*, 89–101.

Knox, J.H. and Pryde, A. (1975) J. Chromatogr. *112*, 171–188.

Knox, J.H. and Vasvari, G. (1973) J. Chromatogr. *83*, 181–189.

Konnan, A.W. (1980) J. Lipid Res. *21*, 780–784.

Kono, A., Hara, Y., Eguchi, S. and Tanaka, M. (1979) J. Chromatogr. *164*, 404–406.

Kopaciewicz, W. and Regnier, F.E. (1982) Anal. Biochem. *126*, 8–16.

Kringe, K.P., Neidhart, B. and Lippman, C. (1983) Pract. Aspects Mod. High Perf. Liq. Chromatogr. *81*, 241–273.

Krinsky, N.I. and Wecankiwar, S. (1984) Methods Enzymol. *105*, 155–170.

Krstulovic, A.M., Hartwick, R.A., Brown, P.R. and Lohse, K. (1978) J. Chromatogr. *158*, 365–376.

Krstulovic, A.M., Colin, H. and Guiochon, G. (1982) Anal. Chem. *54*, 2438–2443.

Kuhn, R., Winterstein, A. and Lederer, E. (1931) Hoppe-Seyler's Z. Physiol. Chem. *197*, 141–160.

Kundu, S.K. and Scott, D.D. (1982) J. Chromatogr. *232*, 19–27.

Kuo, K.C., McCune, R.A. and Gehrke, C.W. (1980) Nucleic Acids Res. *8*, 4763–4776.

Kurganov, A.A., Tevlin, A.B. and Davankov, V.A. (1983) J. Chromatogr. *261*, 223–228.

Lamblin, G., Klein, A., Boersma, A., Nasir-ud-din, H. and Roussel, P. (1983) Carbohydr. Res. *118*, C1–C4.

Langenberg, J.P. and Tjaden, U.R. (1984) J. Chromatogr. *305*, 61–72.

Lapeyre, J.-N. and Becker, F.F. (1979) Biochem. Biophys. Res. Commun. *87*, 698–705.

Larmann, J.P., DeStefano, J.J., Goldberg, A.P., Stout, R.W., Snyder, L.R. and Stadalius, M.A. (1983) J. Chromatogr. *255*, 163–189.

Larsson, P.O. (1984) Methods Enzymol. *104*, 212–223.

Lawrence, J.F. and Frei, R.W. (1976) Chemical Derivatisation in Liquid Chromatography. Elsevier, Amsterdam.

Lebelle, M.J., Lauriant, G. and Wilson, W.L. (1980) J. Liq. Chromatogr. *3*, 1573–1580.

Lecaillon, J.B., Rouan, M.C., Souppart, C., Febvre, N. and Juge, F. (1982) J. Chromatogr. *228*, 257–267.

Lee, D.P. (1982) J. Chromatogr. Sci. *20*, 203–208.

Lee, G.J.-L., Liu, D.-W., Pav, J.W. and Tieckelmann, H. (1981) J. Chromatogr. *212*, 65–73.

Lee, W.M.F., Westrick, M.A. and Macher, M.A. (1982) Biochim. Biophys. Acta *712*, 498–504.

Lefevre, M.D., De Leenheer, A.P. and Claeys, A.E. (1979) J. Chromatogr. *186*, 749–762.

Lehmann, A. and Wittmann-Liebold. B. (1984) FEBS Lett. *176*, 360–364.

Leonard, N.J. and Tolman, G.L. (1975) Ann. N.Y. Acad. Sci. *255*, 43–58.

Lesec, J. (1985) J. Liq. Chromatogr. *8*, 875–923.

Lewis, R.V., Fallon, A., Gibson, K.D., Stein, S. and Udenfriend, S. (1980) Anal. Biochem. *104*, 153–159.

Lieber, L.F., Kuo, J., Krigel, R., Pille, E. and Silber, R. (1981) Anal. Biochem. *118*, 53–60.

Lietzke, M.H., Stoughton, R.W. and Fuoss, R.M. (1968) Proc. Natl. Acad. Sci. U.S.A. *50*, 39–45.

Lim, C.K. (1982) J. Liq. Chromatogr. *5* (Suppl. 2), 305–318.

Lim, K.L., Young, R.W. and Driskell, J.A. (1980) J. Chromatogr. *188*, 285–292.

Lin, J.K. (1984) CRC Handbook for the Separation of Amino Acids, Peptides and Proteins, *1*, 359–366.

Lin, J.-T. and Heftmann, E. (1981) J. Chromatogr. *212*, 239–244.

Lin, J.-T., Heftmann, E. and Hunter, I.R. (1980) J. Chromatogr. *190*, 169–174.

Lindner, W., Le Page, J.N., Davies, G., Seitz, D.E. and Karger, B.L. (1980) J. Chromatogr. *185*, 323–329.

Lindqvist, B., Sjogren, I. and Nordin, R. (1974) J. Lipid Res. *15*, 65–73.

Linz, U. (1983) J. Chromatogr. *260*, 161–163.

Little, C.J., Dale, A.D. and Whatley, J.A. (1979a) J. Chromatogr. *171*, 431–434.

Little, C.J., Whatley, J.A., Dale, A.D. and Evans, M.B. (1979b) J. Chromatogr. *171*, 435–438.

Lochmuller, C.H. and Hill, W.B. Jr., (1983) J. Chromatogr. *264*, 215–222.

Lochmuller, C.H. and Wilder, D.R. (1979) J. Chromatogr. Sci. *17*, 574–579.

Lochmuller, C.H., Ryall, R.R. and Amoss, C.W. (1979) J. Chromatogr. *178*, 298–301.

Lochmuller, C.H., Hangae, H.H. and Wilder, D.R. (1981) J. Chromatogr. Sci. *19*, 130–136.

Loo, J.C.K., Butterfield, A.G., Moffatt, J. and Jordan, N. (1977) J. Chromatogr. *143*, 275–280.

Low, G.K.C., Haddad, P.R. and Duffield, A.M. (1983) J. Chromatogr. *261*, 345–356.

Lowe, C.R. (1979) An Introduction to Affinity Chromatography. Laboratory Techniques in Biochemistry and Molecular Biology Vol. 7, Part 2. Elsevier, Amsterdam.

Lowe, C.R., Glad, M., Larsson, P.-O., Ohlson, S., Small, D.A.P., Atkinson, A. and Mossbach, K. (1981) J. Chromatogr. *215*, 303–316.

Luiken, J., Van den Zee, R. and Welling, G.W. (1984) J. Chromatogr. *284*, 482–486.

Lundanes, E., Dahl, J. and Greibrokk, T. (1983) J. Chromatogr. Sci. *21*, 235–240.

Macrae, R. (ed.) (1982) HPLC in Food Analysis. Academic Press, New York NY. p. 154.

Macrae. R. and Dick, J. (1981) J. Chromatogr. *210*, 138–145.

Mahoney, W.C. and Hermodson, M.A. (1980) J. Biol. Chem. *255*, 11199–11203.

Mai, J., Goswami, S.K., Bruckner, G. and Kinsella, J.E. (1982) J. Chromatogr. *230*, 15–28.

Maitra, S.K., Yoshikawa, T.T., Hansen, J.L., Nilsson-Ehle, I., Palin, W., Schotz, M.C. and Guze, L.B. (1977) Clin. Chem. *23*, 2275–2278.

Majors, R.E. (1972) Anal. Chem. *44*, 1722–1726.

Majors, R.E. (1980) J. Chromatogr. Sci. *18*, 571–575.

Marai, L., Myher, J.J. and Kuksis, A. (1983) Can. J. Biochem. Cell Biol. *61*, 840–849.

Margosis, M. (1982) J. Chromatogr. *236*, 469–480.

Marples, J. and Oates, M.D.G. (1982) J. Antimicrob. Chemother. *10*, 311–318.

Martin, A.J.P. (1969) In: Historical Background in Gas Chromatography in Biology and Medicine (1969 Ciba Foundation Symposium) (Porter, R., ed.) Churchill, London. pp. 2–10.

Martin, A.J.P. and Synge, R.L.M. (1941) Biochem. J. *35*, 1358–1368.

Martinez-Valdez, H., Kothari, R.M., Hershey, H.V. and Taylor, M.W. (1982) J. Chromatogr. *247*, 307–314.

Maskarinec, M.P., Vargo, J.D. and Sepaniak, M.J. (1983) J. Chromatogr. *261*, 245–251.

Masters, L.G. and Leyden, D.E. (1978) Anal. Chim. Acta 98, 9–15.
Matlin, S.A., Tito-Lloret, A., Longh, W.J., Bryan, D.G., Browne, T. and Mehani, S. (1981) J. High Res. Chromatogr. Commun. 4, 81–83.
Maybaum, J., Klein, F.K. and Sadee, W. (1980) J. Chromatogr. 188, 149–158.
Mays, D.L., Van Apeldsorn, R.J. and Laubaca, R.G. (1976) J. Chromatogr. 120, 93–102.
McCann, M., Purnell, H. and Wellington, C.A. (1982) Proc. Faraday Soc. 15, 1–91.
McCluer, R.H. and Evans, J.E. (1973) J. Lipid Res. 14, 611–617.
McCluer, R.H. and Evans, J.E. (1976) J. Lipid Res. 17, 412–418.
McFarland, G.D. and Borer, P.N. (1979) Nucleic Acids Res. 7, 1067–1080.
McGinnis, G.D. and Fang, P. (1978) J. Chromatogr. 153, 107–114.
McGuffin, V.L. and Novotny, M. (1983) J. Chromatogr. 255, 381–393.
McLaughlin, L.W. and Romanuik, E. (1982) Anal Biochem. 124, 37–44.
McNair, H.M. (1984) J. Chromatogr. Sci. 22, 521–535.
McNair, H.M. and Chandler, C.D. (1976) J. Chromatogr. Sci. 4, 477–483.
Meek, J.C. and Rosetti, Z.L. (1981) J. Chromatogr. 211, 15–28.
Mefford, I.N. (1981) J. Neurosci. Methods 3, 207–224.
Melander, W.R. and Horvath, C.S. (1979) J. Chromatogr. 185, 129–152.
Melander, W.R. and Horvath, C.S. (1980a) High Perf. Liq. Chromatogr. 2, 113–319.
Melander, W.R. and Horvath, C.S. (1980b) J. Chromatogr. 201, 211–218.
Melander, W.R., Chen, B.-K. and Horvath, C. (1979a) J. Chromatogr. 185, 99–109.
Melander, W.R., Nahum, A. and Horvath, C. (1979b) J. Chromatogr. 185, 129–152.
Melander, W.R., Stoveken, J. and Horvath, C. (1979c) J. Chromatogr. 185, 111–127.
Mellis, S.J. and Baenziger, J.U. (1981) Anal. Biochem. 114, 276–280.
Mellis, S.J. and Baenziger, J.U. (1983) Anal. Biochem. 134, 442–449.
Meredith, S. (1984) J. Biol. Chem. 259, 11682–11685.
Merritt, M.V. and Bronson, G.E. (1977) Anal. Biochem. 80, 392–400.
Metz, S.A., Hall, M.E., Harper, T.W. and Murphy, R.C. (1982) J. Chromatogr. 233, 193–201.
Miller, R. (1982) Anal. Chem. 54, 1742–1746.
Miller, R.D. and Neuss, N. (1983) Handbook Exp. Pharmacol. 67, 301–328.
Miller, R.A., Bussell, N.E. and Ricketts, C. (1978) J. Liq. Chromatogr. 1, 291–304.
Miller, A.A., Benvenuto, J.A. and Loo, T.L. (1982) J. Chromatogr. 228, 165–176.
Miyazaki, K., Ohtani, K., Sunada, K. and Arita, T. (1983) J. Chromatogr. 276, 478–482.
Molko, D., Derbyshire, R., Guy, A., Roget, A., Teoule, R. and Boucherle, A. (1981) J. Chromatogr. 206, 493–500.
Molnar, I. and Horvath, C. (1976) Clin. Chem. 22, 1497–1501.
Montelaro, R.C., West, M., and Issel, C.J. (1981) Anal. Biochem. 114, 398–406.
Moonen, P., Klok, G. and Keirse, M.J.N.C. (1983) Prostaglandins 26, 797–803.
Mori, K. (1974) Jpn. J. Ind. Health 16, 490–493.
Mori, S. (1979) J. Chromatogr. 174, 23–33.
Moro, E., Bellotti, V., Jannuzzo, M.G., Stegnjaich, S. and Valzelli, G. (1983) J. Chromatogr. 274, 281–287.
Morrison, W.H., Lou, M.F. and Hamilton, P.B. (1976) Anal. Biochem. 71, 415–425.
Mowery, R.A. (1985) J. Chromatogr. Sci. 23, 22–29.
Moyer, T.P. and Jiang, N.-S. (1978) J. Chromatogr. 153, 365–372.
Mrochek, J.E., Dinsmore, S.R. and Waalkes, T.P. (1975) Clin. Chem. 21, 1314–1322.
Muller, A.J. and Carr, P.W. (1984) J. Chromatogr. 284, 33–51.

Murphy, R.C., Hammerstrom, S. and Samuelsson, B. (1976) Proc. Natl. Acad. Sci. U.S.A. 76, 4275–4279.

Musey, P.I., Collins, D.C. and Preedy, J.R.K. (1978) Steroids 31, 583–592.

Myher, J.J., Kuksis, A., Marai, L. and Manganaro, F. (1984) J. Chromatogr. 283, 289–301.

Nachtmann, F. and Budna, K.W. (1977) J. Chromatogr. 136, 279–287.

Nahum, A. and Horvath, C. (1981) J. Chromatogr. 203, 53–63.

Naider, F., Huchital, M. and Becker, J.M. (1983) Biopolymers 22, 1401–1407.

Nakagawa, T., Shibukawa, A. and Uno, T. (1983) J. Chromatogr. 254, 27–34.

Nakamura, R., Yamaguchi, T., Sekine, Y. and Hashimoto, M. (1983) J. Chromatogr. 278, 321–328.

Nakano, K., Assenza, S.P. and Brown, P.R. (1982) J. Chromatogr. 233, 51–60.

Namambara, T., Ikegawa, S., Hasegawa, M. and Goto, J. (1974) Anal. Chim. Acta 101, 111–115.

Nebinger, P., Koel, M., Franz, A. and Werries, E. (1983) J. Chromatogr. 265, 19–25.

Newman, P.J. and Kahn, R.A. (1983) Anal. Biochem. 132, 215–218.

Newton, C.R., Greene, A.R., Heathcliffe, G.R., Atkinson, T.C., Holland, D., Markham, A.F. and Edge, M.D. (1983) Anal. Biochem. 129, 22–30.

Nice, E.C. and O'Hare, M.J. (1979) J. Chromatogr. 162, 401–407.

Nielsen, P., Rauschenbach, D. and Bacher, A. (1983) Anal. Biochem. 130, 359–368.

Nilsson, B. (1983) J. Chromatogr. 276, 413–417.

Nilsson, K. and Larsson, P.O. (1983) Anal. Biochem. 134, 60–72.

Nilsson-Ehle, I. (1983) J. Liq. Chromatogr. 6, 251–293.

Nilsson-Ehle, I., Yoshikawa, T.T., Schotz, M.C. and Guze, L.B. (1976) Antimicrob. Agents Chemother. 9, 754–760.

Nissinen, E. (1980) Anal. Biochem. 106, 497–505.

Nordin, P. (1983) Anal. Biochem. 131, 492–498.

Novotny, M. (1981) Anal. Chem. 53, 1294A–1308A.

O'Hare, M.J. and Nice, E.C. (1982) Chromatogr. Sci. 16, 277–322.

O'Hare, M.J., Capp, M.W., Nice, E.C., Cooke, N.H.C. and Archer, B.G. (1982) Anal. Biochem. 126, 17–28.

Ohlson, S., Hanson, L., Larsson, P.O. and Mossbach, K. (1978) FEBS Lett. 93, 5–7.

Ohno, Y., Okasaki, M. and Hara, I. (1981) J. Biochem. 89, 1675–1681.

Oi, N., Wagase, M. and Doi, T. (1983) J. Chromatogr. 265, 117–126.

Okamoto, K.-I., Ishida, Y. and Asai, K. (1978) J. Chromatogr. 167, 205–217.

Ondrus, M.G., Wenzel, J. and Zimmerman, G.L. (1983) J. Chem. Ed. 60, 776–778.

Ordemann, D.M. and Walton, H.F. (1976) Anal. Chem. 48, 1728–1730.

Osborne, D.J., Peters, B.J. and Meade, C.J. (1983) Prostaglandins 26, 817–832.

Otto, M. and Wegscheider, W. (1983) J. Chromatogr. 258, 11–22.

Ozcimder, M. and Hammers, W.E. (1980) J. Chromatogr. 187, 307–317.

Paart, E. and Samuelson, O. (1973) J. Chromatogr. 85, 93–99.

Palmer, J.K. (1975) Anal. Lett. 8, 215–224.

Pandakker, J.E. and Groenedijk, G.W.T. (1979) J. Chromatogr. 168, 125–131.

Papp, E. and Vigh, G. (1983) J. Chromatogr. 259, 49–58.

Parente, J.P., Strecker, G., Leroy, Y., Montreuil, J. and Fournet, B. (1982) J. Chromatogr. 249, 199–204.

Parris, N.A. (1978) J. Chromatogr. 157, 161–170.

Parrish, D.B. (1980) CRC Crit. Rev. Food Sci. Nutr. 13, 161–170.

Pascal, R.A., Farris, C.L. and Schroepfer, G.J. (1980) Anal. Biochem. 101, 15–22.

Paton, R.D., McGillivray, A.I., Speir, T.F., Whittle, M.J., Whitfield, C.R. and Logan, R.W. (1983) Clin. Chim. Acta *133*, 97–110.

Patthy, M. and Gyenge, R. (1984) J. Chromatogr. *286*, 217–221.

Patton, G.M., Fasulo, J.M. and Robins, S.J. (1982) J. Lipid Res. *23*, 190–195.

Pauliukonis, L.T., Musson, D.G. and Bayne, W.F. (1984) J. Pharm. Sci. *73*, 99–102.

Payne, S.M. and Ames, B.N. (1982) Anal. Biochem. *123*, 151–161.

Payne-Wahl, K., Plattner, R.D., Spencer, G.F. and Kleiman, R. (1979) Lipids *14*, 601–606.

Pearson, J.D. and Regnier, F.E. (1983) J. Chromatogr. *255*, 137–149.

Pearson, J.D., Lin, T.N. and Regnier, F.E. (1982) Anal. Biochem. *124*, 217–230.

Pei, P.T.S., Henly, R.S. and Ramachandran, S. (1975) Lipids *10*, 152–155.

Pellinen, J. and Salkinoja-Salonen, M. (1985) J. Chromatogr. *328*, 299–308.

Peng, G.W., Gadalla, M.A.F., Peng, A., Smith, V. and Chiou, W.L. (1977a) Clin. Chem. *23*, 1838–1844.

Peng, G.W., Jackson, G.G. and Chiou, W.L. (1977b) Antimicrob. Agents Chemother. *12*, 707–712.

Perkins, E.G., Hendren, D.J., Bauer, J.E. and El-Hamdy, A.H. (1981) Lipids *16*, 609–613.

Perrett, D. (1976) J. Chromatogr. *124*, 187–196.

Perrin, D.D., Dempsey, B. and Sarjeant, E.P. (eds.) (1980) pKa Prediction for Organic Acids and Bases. Chapman and Hall, London.

Peters, S.P., Schulman, E.S., Liu, M.C., Hayes, E.C. and Lichtenstein, L.M. (1983) J. Immunol. Methods *64*, 335–343.

Pieroni, J.P., Lee, W.H. and Wong, P.Y.K. (1982) J. Chromatogr. *230*, 115–120.

Plattner, R.D. and Payne-Wahl, K. (1979) Lipids *14*, 152–153.

Plattner, R.D., Spencer, G.E. and Kleiman, R. (1977) J. Am. Oil Chem. Soc. *54*, 511–515.

Plattner, R.D., Wade, K. and Kleiman, R. (1978) J. Am. Oil. Chem. Soc. *55*, 381–383.

Pogolotti, A.L. Jr. and Santi, D.V. (1982) Anal. Biochem. *126*, 335–345.

Polizer, I.R., Griffin, G.W., Dowty, B.J. and Laseter, L. (1973) Anal. Lett. *6*, 539–540.

Poppe, H. and Kraak, J.C. (1983) J. Chromatogr. *255*, 395–414.

Porath, J. (1981) J. Chromatogr. *218*, 241–259.

Porath, J. and Flodin, P. (1959) Nature *183*, 1657–1659.

Porter, N.A., Wolf, R.A. and Nixon, J.R. (1979) Lipids *14*, 20–24.

Powell, W.S. (1982) Methods Enzymol. *86*, 530–543.

Prescott, W.R., Boyd, B.K. and Seaton, J.F. (1982) J. Chromatogr. *234*, 513–516.

Preston, M.R. (1983) J. Chromatogr. *275*, 178–182.

Purdy, R.H., Durocher, C.K., Moore, P.H. and Rao, P.N. (1982) Chromatogr. Sci. *16*, 81–104.

Rabel, F.M. (1980) J. Chromatogr. Sci. *18*, 394–396.

Rabel, F.M., Caputo, A.G. and Butts, E.T. (1976) J. Chromatogr. *126*, 731–740.

Radmark, O., Malmsten, C., Samuelsson, B., Goto, G., Marfat, A. and Corey, E.J. (1980) J. Biol. Chem. *255*, 11828–11831.

Radus, T.P. and Gyr, G. (1983) J. Pharm. Sci. *72*, 221–224.

Ramos, D.L. and Schoffstall, A.M. (1983) J. Chromatogr. *261*, 83–93.

Randerath, E., Yu, C.-T. and Randerath, K. (1972) Anal. Biochem. *48*, 172–198.

Reardon, G.E., Caldarella, A.M. and Canalis, E. (1979) Clin. Chem. *25*, 122–126.

Reed, L.S. and Archer, M.C. (1976) J. Chromatogr. *121*, 100–109.

Refshauge, C., Kissinger, P.T., Dreiling, R., Blank, L., Freeman, R. and Adams, R.N. (1974) Life Sci. *14*, 311–314.

Regnier, F.E. and Gooding, K.M. (1980) Anal. Biochem. *103*, 1–25.
Regnier, F.E. and Noel, R. (1976) J. Chromatogr. Sci. *14*, 316–320.
Richmond, M.L., Brandao, S.C.C., Gray, J.I., Markakis, P. and Stine, C.M. (1981) J. Agric. Food Chem. *29*, 4–7.
Richmond, M.L., Barfuss, D.L., Harte, B.R., Gray, J.I. and Stine, C.M. (1982) J. Dairy Sci. *65*, 1394–1400.
Riley, C.M., Tomlinson, E. and Jeffries, T.M. (1979) J. Chromatogr. *185*, 197–224.
Robinson, Y., Gerry, N. and Vose, C.W. (1985) J. Chromatogr. *338*, 219–224.
Rogers, M.E., Adlard, M.W., Saunders, G. and Holt, G. (1983) J. Chromatogr. *257*, 91–150.
Rogozhin, S.V. and Davankov, V.A. (1971) J. Chem. Soc. D Chem. Commun. *490*.
Roos, R.W. (1976) J. Chromatogr. Sci. *14*, 505–512.
Rosenfelder, G., Chang, J.-Y. and Braun, D.G. (1983) J. Chromatogr. *272*, 21–27.
Roston, D.A. and Kissinger, P.T. (1981) Anal. Chem. *53*, 1695–1699.
Roston, D.A., Shoup, R.E. and Kissinger, P.T. (1982) Anal. Chem. *54*, 1417A–1434A.
Roumeliotis, P. and Unger, K.K. (1979) J. Chromatogr. *185*, 445–450.
Roy, S.K., Weber, D.V. and McGregor, W.C. (1984) J. Chromatogr. *303*, 225–228.
Rubinstein, M. Rubinstein, S., Familletti, P.C., Miller, R.S., Waldman, A.A. and Pestka, S. (1979) Proc. Natl. Acad. Sci. U.S.A. *76*, 640–645.
Rubinstein, M., Levy, W.P., Moschera, J.A., Lai, C.Y., Hershberg, R., Bartlett, R. and Pestka, S. (1980) Arch. Biochem. Biophys. *210*, 307–318.
Ryba, M. and Beranek, J. (1981) J. Chromatogr. *211*, 337–346.
Sakodynskii, K. (1970) J. Chromatogr. *49*, 2–17.
Salib, A.G., Frappier, F., Guillem, G. and Marquet, A. (1979) Biochem. Biophys. Res. Commun. *88*, 312–318.
Sampson, R.L. (1977) Am. Lab. *9*, 109–113.
Sassenfeld, H.M. and Brewer, S.J. (1984) Biotechnology 2, 76–81.
Schill, G. (1974) In: Ion Exchange and Solvent Extraction, Vol. 6 (Marinsky, J.A. and Marcus, Y., eds.). Marcel Dekker, New York, NY. pp. 1–30.
Schlabach, T.D. and Wehr, T.C. (1982) Anal. Biochem. *127*, 222–233.
Schmidt, G.J. (1982) Chromatogr. Sci. *16*, 45–72.
Schmidt, G.J., Vandemark, F.L. and Slavin, W. (1978) Anal. Biochem. *91*, 636–645.
Schmitt, J.A., Henry, R.A., Williams, R.C. and Dickman, J.F. (1971) J. Chromatogr. Sci. *9*, 645–649.
Schoenmakers, P.J., Biliet, H.A. and De Galan, L. (1979) J. Chromatogr. *185*, 179–185.
Scholfield, C.R. (1979) J. Am. Oil Chem. Soc., *56*, 510–516.
Scholten, A.H.M.T., Brinkman, U.A. and Frei, R.W. (1981) J. Chromatogr. *205*, 229–237.
Schott, H., Meyer, H.D. and Bayer, E. (1983) J. Chromatogr. *280*, 297–311.
Schusler-Van Hees, M.T.I.W. and Beijersbergen-Van Henegouwen, G. (1980) J. Chromatogr. *196*, 101–108.
Schuster, R. (1980) Anal. Chem. *52*, 617–620.
Schwedt, G. (1979) Angew. Chem. Int. Ed. Engl. *18*, 180–186.
Schwedt, G. and Bussemas, H. (1976) Chromatographia *9*, 17–21.
Schwedt, G. and Bussemas, H. (1977) Z. Anal. Chem. *285*, 381–385.
Schwedt, G., Bussemas, H.H. and Lippmann, C.H. (1977) J. Chromatogr. *143*, 259–266.
Scopes, R. (1982) Protein Purification: Principles and Practice. Springer-Verlag, New York NY.
Scott, C.D. (1971) Modern Practice of Liquid Chromatography (Kirkland, J.J., ed.) Wiley Interscience, New York NY.

Scott, R.P.W. (1977) J. Chromatogr. *11*.

Scott, R.P.W. (1982) Techniques of Liquid Chromatography (Simpson, C.F., ed.) Wiley-Heden, Chichester.

Scott, N.R. and Dixon, P.F. (1979) J. Chromatogr. *164*, 29–34.

Scott, I.M. and Horgan, R. (1982) J. Chromatogr. *237*, 311–315.

Scott, R.P.W. and Kucera, P. (1973) Anal. Chem. *45*, 749–752.

Scott, R.P.W. and Kucera, P. (1978) J. Chromatogr. *149*, 93–97.

Scott, R.P.W. and Kucera, P. (1979) J. Chromatogr. *185*, 27–41.

Scratchley, G.A., Masoud, A.N., Stohs, S.J. and Wingard, D.W. (1979) J. Chromatogr. *169*, 313–319.

Seki, T. (1978) J. Chromatogr. *155*, 415–421.

Seki, T. and Yamaguchi, Y. (1983) J. Liq. Chromatogr. *6*, 1131–1138.

Seki, T. and Yamaguchi, Y. (1984a) J. Chromatogr. *287*, 407–412.

Seki, T. and Yamaguchi, Y. (1984b) J. Chromatogr. *305*, 188–193.

Seliger, H., Bach, T.C., Gortz, H.H., Happ, E., Holupirek, M., Seemann-Preising, B., Schiebel, H.M. and Schulten, H.R. (1982) J. Chromatogr. *253*, 65–79.

Senftleber, F., Bowling, D. and Stahr, M.S. (1983) Anal. Chem. *55*, 810–812.

Shearer, M.J. (1983) Adv. Chromatogr. *21*, 243–249.

Shimada, K., Tanaka, M. and Nambara, T. (1979) J. Chromatogr. *178*, 350–354.

Shimada, K., Tanaka, M. and Nambara, T. (1980) Anal. Lett. *13*, 1129–1133.

Shimada, K., Tanaka, M. and Nambara, T. (1981) J. Chromatogr. *307*, 23–28.

Shimada, K., Tanaka, T. and Nambara, T. (1984) J. Chromatogr. *223*, 33–39.

Shine, J., Fettes, I., Lan, N.C.Y., Roberts, J.L. and Baxter, J.D. (1980) Nature *285*, 456–461.

Shoup, R.E. and Kissinger, P.T. (1977) Clin. Chem. *23*, 1268–1274.

Silver, M. (1979) Methods Enzymol. *62*, 135–137.

Sinanoglu, O. and Abdulnur, S. (1965) Fed. Proc. *243*, 12–18.

Sinkule, J.A. and Evans, W.E. (1983) J. Chromatogr. *274*, 87–93.

Sisco, W.R. and Gilpin, R.K. (1980) J. Chromatogr. Sci. *18*, 41–45.

Skelly, N.E. (1982) Anal. Chem. *54*, 712–715.

Slikker, W., Lipe, G.W. and Newport, G.D. (1981) J. Chromatogr. *224*, 205–219.

Slikker, W., Althaus, Z.R., Rowland, J.M., Hill, D.E. and Hendrickx, A.G. (1982) J. Pharm. Exp. Ther. *223*, 368–374.

Small, D.A.P., Atkinson, T. and Lowe, C.R. (1981) J. Chromatogr. *216*, 175–183.

Smith, M. and Jungalwala, F.B. (1981) J. Lipid Res. *22*, 697–704.

Smith, E.C., Jones, A.D. and Hammond, E.W. (1980) J. Chromatogr. *188*, 205–212.

Smolenski, K.A., Avery, N.C. and Light, N.D. (1983a) Biochem. J. *213*, 525–532.

Smolenski, K.A., Fallon, A., Light, N.D. and Bailey, A.J. (1983b) Biosci. Rep. *3*, 93–100.

Smolenski, K.A., Fallon, A. and Light, N.D. (1984) J. Chromatogr. *287*, 29–44.

Snyder, L.R. (1968) Principles of Adsorption Chromatography. Dekker, New York NY.

Snyder, L.R. (1974) Anal. Chem. *46*, 1384–1393.

Snyder, L.R. (1976) J. Chromatogr. *125*, 287–306.

Snyder, L.R. (1978) J. Chromatogr. *149*, 653–668.

Snyder, L.R. (1979) J. Chromatogr. *179*, 167–172.

Snyder, L.R. (1980) J. Chromatogr. *184*, 363–414.

Snyder, L.R. (1983) High Performance Liquid Chromatography: Advances and Perspectives. Academic Press, New York NY.

Snyder, L.R. and Kirkland, J.J. (1979) Introduction to Modern Liquid Chromatography. Wiley-Interscience, New York NY.

Snyder, L.R. and Poppe, H. (1980) J. Chromatogr. *184*, 363–371.
Snyder, L.R. and Schunk, T.C. (1982) Anal. Chem. *54*, 1764–1772.
Soczewinski, E. (1968) Anal. Chem. *41*, 179–183.
Soczewinski, E. and Waksmundzka-Hajnus, M. (1980) J. Liq. Chromatogr. *3*, 1625–1636.
Sonnefeld, W.J., Zoller, W.H., May, W.E. and Wise, S.A. (1982) Anal. Chem. *54*, 723–727.
Spackman, D.H., Stein, W.H. and Moore, S. (1950) Anal. Chem. *30*, 1190–1205.
Speek, A.J., Schriver, J. and Schreurs, W.H.P. (1984) J. Chromatogr. *305*, 53–60.
Sportsman, R.T. and Wilson, G.S. (1980) Anal. Chem. *52*, 2013–2017.
Stancher, B. and Zonta, F. (1983) J. Chromatogr. *286*, 93–98.
Stein, S. and Brink, L. (1981) Methods Enzymol. *79*, 120–121.
Stewart, P.S., Bailey, P.A. and Bevan, J.L. (1983) J. Chromatogr. *282*, 589–593.
Stillman, R. and Ma, T.S. (1974) Michrochim. Acta *82*, 641–648.
Stokes, R.H. and Walton, H.F. (1954) J. Am. Chem. Soc. *76*, 3327–3330.
Stout, R.W., De Stefano, J.J. and Snyder, L.R. (1983) J. Chromatogr. *261*, 189–212.
Stranahan, J.J. and Deming, S.N. (1982) Anal Chem. *54*, 2251–2256.
Strickler, M.P., Travis, R.W. and Olenick, J.G. (1982) J. Liq. Chromatogr. *5*, 1933–1939.
Stulik, K. and Pacakova, V. (1982) CRC Crit. Rev. Anal. Chem. *14*, 297–351.
Sugden, K., Hunter, C., Lloyd-Jones, J.G. (1980) J. Chromatogr. *192*, 228–235.
Sugita, M., Iwamori, N., Evans, J.E., McCluer, R.H., Dulaney, J.T. and Moser, H.W. (1979) J. Lipid Res. *15*, 223–227.
Sulkowski, E., Vastola, K., Oieszek, D. and Von Muenchausen, W. (1982) Anal. Chem. Symp. Ser. Vol. 9, Affinity Chromatography and Related Techniques (Gribuan, T.C.J., Vissen, J. and Nivard, R.F.J., eds.) Elsevier, Amsterdam, pp. 313–322.
Sullivan, R.C., Shing, V.W., D'Amore, P.A. and Klagsbrun, M. (1983) J. Chromatogr. *266*, 301–311.
Suzuki, A., Kundu, S.K. and Marcus, D.M. (1980) J. Lipid Res. *21*, 473–477.
Svensson, L., Sisfontes, L., Nyborg, G. and Blomstrand, J. (1982) Lipids *17*, 50–55.
Sztarickskai, F., Borda, J., Puskas, M. and Bognar, R. (1983) J. Antibiotics *36*, 1691–1698.
Tanaka, N., Tokuda, Y., Iwaguchi, K. and Araki, M. (1982) J. Chromatogr. *239*, 761–772.
Tang, M. and Deming, S.N. (1983) Anal. Chem. *55*, 425–428.
Tapuhi, Y., Schmidt, D.E., Linder, W. and Karger, B.L. (1981) Anal. Biochem. *52*, 595–601.
Tartivita, K.A., Sciarello, J.P. and Rudy, B.C. (1976) J. Pharm. Sci. *65*, 1024–1029.
Taylor, J.T., Knotts, J.G. and Schmidt, G.J. (1980) Clin. Chem. *26*, 130–132.
Taylor, R.B., Reid, R., Kendle, K.E., Geddes, C. and Curle, P.F. (1983) J. Chromatogr. *277*, 101–114.
Tempst, P., Hunkapiller, M.W. and Hood, L.E. (1984) Anal. Biochem. *137*, 188–195.
Thiem, J., Schwenter, J., Karl, H., Sievers, A. and Reimer, J. (1978) J. Chromatogr. *155*, 107–118.
Thomas, J.-P., Brun, A. and Bonnine, J.-P. (1979) J. Chromatogr. *172*, 107–130.
Thompson, R.M. (1978) J. Chromatogr. *166*, 201–212.
Tilly Melin, A., Ljungcrantz, M. and Schill, G. (1979) J. Chromatogr. *185*, 225–239.
Tjaden, U.R., Krol. J.H., Van Hoeven, R.P., Oomen-Meulemans, E.P.M. and Emmelot, P. (1977) J. Chromatogr. *136*, 233–243.
Todoriki, H., Hayashi, T. and Naruse, H. (1983) J. Chromatogr. *276*, 45–54.

Touchstone, J.C. and Wortman, W. (1973) J. Chromatogr. *76*, 244–247.

Traylor, T.D., Koontz, D.A. and Hogan, E.L. (1983) J. Chromatogr. *272*, 9–20.

Trefz, F.K., Byrd, D.J. and Kochen, W. (1975) J. Chromatogr. *107*, 181–189.

Tsao, C.S. and Salin, S.L. (1981) J. Chromatogr. *224*, 477–486.

Tscherne, R.J. and Capitano, G. (1977) J. Chromatogr. *136*, 337–341.

Tsimidou, M. and Macrae, R. (1984) J. Chromatogr. *285*, 178–181.

Tsuchiya, H., Takagi, N., Hayashi, T. and Naruse, H. (1981) Rinsho Kagaku Shimpojumu *21*, 43–47.

Tsuki, K. and Binns, R.B. (1982) J. Chromatogr. *253*, 227–236.

Tsuruta, Y., Kohashi, K., Ishida, S. and Ohkura, Y. (1984) J. Chromatogr. *309*, 309–315.

Turk, J., Weiss, S.J., Davis, J.E. and Needleman, P. (1978) Prostaglandins *16*, 291–309.

Twitchett, P.J. and Moffat, A.C. (1975) J. Chromatogr. *111*, 149–157.

Udenfriend, S., Stein, S., Bohlen, P., Dairman, W., Leimgruber, W. and Weigle, W. (1972) Science *178*, 871–873.

Ui, N. (1979) Anal. Biochem. *97*, 65–73.

Ui, N. (1981) J. Chromatogr. *215*, 289–292.

Ulick, S., Ramirez, L.C. and New, M.I. (1977) J. Clin. Endocrinol. *44*, 799–802.

Ullman, M.D. and McCluer, R.H. (1977) J. Lipid Res. *18*, 371–378.

Ullman, M.D. and McCluer, R.H. (1978) J. Lipid Res. *19*, 910–913.

Unger, K. and Messer, W. (1978) J. Chromatogr. *149*, 1–9.

Unger, K., Becker, N. and Roumeliotis, P. (1976) J. Chromatogr. *125*, 115–127.

Van Bockstaele, M., Dillen, L., Claeys, M. and De Potter, W.P. (1983) J. Chromatogr. *275*, 11–20.

Van den Berg, J.H.M., Mol, C.R., Deelder, R.S. and Thijssen, J.H. (1977) Clin. Chim. Acta *78*, 165–172.

Van der Hoeven, J. (1980) J. Chromatogr. *196*, 494–497.

Vanderslice, J. and Maire, C.E. (1980) J. Chromatogr. *196*, 176–179.

Van der Wal, Sj. and Huber, J.F.K. (1974) J. Chromatogr. *102*, 353–374.

Van der Wal, Sj. and Huber, J.F.K. (1977) J. Chromatogr. *135*, 305–321.

Van Lancker, M.A., Nelis, H.J.C. and De Leenheer, A.P. (1983) J. Chromatogr. *254*, 45–52.

Van Olst, H. and Joosten, G.E.H. (1979) J. Liq. Chromatogr. *2*, 111–115.

Van Rollins, M., Aveldano, M.I., Sprecher, H.W. and Horrocks, L.A. (1982) Methods Enzymol. *86*, 518–530.

Veiga, M., Traba, M.P. and Fabregas, J. (1983) J. Antibiotics *36*, 776–783.

Verhaar, L.A.Th. and Kuster, B.F.M. (1981a) J. Chromatogr. *210*, 279–290.

Verhaar, L.A.Th. and Kuster, B.F.M. (1981b) J. Chromatogr. *220*, 313–322.

Verhaar, L.A.Th. and Kuster, B.F.M. (1982) J. Chromatogr. *234*, 57–64.

Verzele, M. and Geeraert, E. (1980) J. Chromatogr. Sci. *18*, 559–570.

Verzele, M. and Muscche, P. (1983) J. Chromatogr. *254*, 117–122.

Vidal-Valverde, C., Martin-Villa, C. and Olmedilla, B. (1982) J. 217–317.

Vigh, Gy. and Varga-Puchony, J. (1980) J. Chromatogr. *196*, 1–9.

Voelter, W. (1985) In: HPLC in Biochemistry. VCH, Weinheim, F.R.G. pp. 217–317.

Voelter, W., Kronbach, T., Zech, K. and Huber, R. (1982) J. Chromatogr. *239*, 475–482.

Vonach, B. and Schomburg, G. (1978) J. Chromatogr. *149*, 417–430.

Von Stetten, O. and Schlett, R. (1983) J. Chromatogr. *254*, 229–235.

Vratny, P., Coupek, J., Vozka, S. and Hostomska, Z. (1983) J. Chromatogr. *254*, 143–155.

Wahlund, K.G. and Beijershen, I. (1982) Anal. Chem. *54*, 128–132.
Wakizaka, A., Kurosaka, K. and Okuhara, E. (1979) J. Chromatogr. *162*, 319–326.
Walker, M.C., Carpenter, B.E. and Cooper, E.L. (1981) J. Pharm. Sci. *70*, 99–101.
Walters, R.R. (1982) J. Chromatogr. *249*, 19–27.
Warner, T.G., Robertson, A.D. and O'Brien, J.S. (1983) Clin. Chim. Acta *127*, 313–326.
Warnock, D.W., Richardson, M.D. and Turner, A. (1982) J. Antimicrob. Chemother. *10*, 467–478.
Warren, C.D., Schmidt, A.S. and Jeanloz, R.W. (1983) Carbohydr. Res. *116*, 171–182.
Watanabe, K. and Arao, Y. (1981) J. Lipid Res. *22*, 1020–1024.
Watkins, W.D. and Peterson, M.B. (1982) Anal. Biochem. *125*, 30–40.
Weatherston, J., MacDonald, L.M., Blake, T., Benn, M.H. and Yuang, Y.Y. (1978) J. Chromatogr. *161*, 347–351.
Weber, S.G. and Purdy, W.C. (1982) Anal. Chem. *54*, 1757–1764.
Webster, H.K. and Whaun, J.M. (1981) J. Chromatogr. *209*, 283–292.
Wehr, T. (1984) Handbook of HPLC for the Separation of Amino Acids, Peptides and Proteins, Vol. 1 (Hancock, W.S., ed.) CRC Press, Florida, pp. 31–57.
Weinshenker, N.M. and Longwell, A. (1972) Prostaglandins *2*, 207–211.
Weinstein, S., Engel, M.H. and Hare, P.E. (1982) Anal. Biochem. *121*, 370–374.
Weiser, E.L., Salotto, A.W., Flach, S.M. and Snyder, L.R. (1984) J. Chromatogr. *303*, 1–12.
Wells, M.J.M. and Clark, C.R. (1981) Anal. Chem. *53*, 1341–1345.
Wells, G.B. and Lester, R.L. (1979) Anal. Biochem. *97*, 184–190.
Wentz, F.E., Marcy, A.D. and Gray, M.J. (1982) J. Chromatogr. Sci. *20*, 349–352.
Werner, W. and Halasz, I. (1980) Chromatographia *13*, 271–273.
Wheals, B.B. (1975) J. Chromatogr. *107*, 402–406.
Wheals, B.B. and White, P.C. (1979) J. Chromatogr. *176*, 421–426.
White, E.R. and Laufer, D.N. (1984) J. Chromatogr. *290*, 187–196.
White, E.R., Carroll, M.A. and Zarembo, J.E. (1977) J. Antibiotics *30*, 811–816.
White, C.A., Corran, P.H. and Kennedy, J.F. (1981) Carbohydr. Res. *87*, 165–173.
White, C.A., Vass, S.W., Kennedy, J.F. and Large, D.G. (1983) J. Chromatogr. *264*, 99–109.
Whitman, W.B. and Wolfe, R.S. (1984) Anal. Biochem. *137*, 261–265.
Whorton, A.R., Carr, K., Smigel, M., Walker, L., Ellis, K. and Oates, J.A. (1979) J. Chromatogr. *163*, 64–71.
Wielders, J.D.M. and Mink, C.J.K. (1983) J. Chromatogr. *277*, 145–156.
Wight, A.W. and van Niekerk, P.J. (1983a) J. Agric. Food Chem. *31*, 282–285.
Wight, A.W. and van Niekerk, P.J. (1983b) Food Chem. *10*, 211–224.
Wightman, R.M., Plotsky, P.M., Strope, E.R., Delore, R.J. and Adams, R.N. (1977) Brain Res. *131*, 345–349.
Williams, R.C., Baker, D.R. and Schmit, J.A. (1973) J. Chromatogr. Sci. *11*, 618–624.
Williams, K.R., L'Italien, J.J., Guggenheimer, R.A., Sillerud, L., Spicer, E., Chase, J. and Konigsberg, W. (1982) Methods in Protein Sequence Analysis. Humana Press, New Jersey.
Wilson, A.E. and Park, B.K. (1983) J. Chromatogr. *277*, 292–299.
Wilson, C.W., Shaw, P.E. and Campbell, C.W. (1982a) J. Sci. Food Agric. *33*, 777–780.
Wilson, K.T., Van Wieringen, E., Klauser, S. and Berthold, M.W. (1982b) J. Chromatogr. *237*, 407–416.
Winzor, D.J. (1985) Affinity Chromatography, a Practical Approach. (Dean, P.G., Johnson, W.S. and Middle, F.A., eds.) IRL Press, Oxford. pp. 149–169.

Wittina, D.P. and Hanley, W.G. (1976) In: GLC and HPLC Analysis of Therapeutic Agents (Tsuji, K. and Monowitch, W., eds.) Dekker, New York NY.

Wong, L.T., Beaubien, A.R. and Pakuts, A.P. (1982) J. Chromatogr. *231*, 145–154.

Wood, P.J. and Siddiqui, I.R. (1971) Carbohydr. Res. *19*, 283–286.

Wood, R., Cummings, L. and Jupille, T. (1980) J. Chromatogr. Sci. *18*, 551–558.

Wynalda, M.A., Lincoln, F.H. and Fitzpatrick, F.A. (1979) J. Chromatogr. *176*, 413–417.

Wynalda, M.A., Morton, D.R., Kelly, R.C. and Fitzpatrick, F.A. (1982) Anal. Chem. *54*, 1079–1082.

Yahara, S., Moser, H.W., Kolodny, E.H. and Kishimoto, Y. (1980) J. Neurochem. *34*, 694–699.

Yali, W.W., Gennard, C.R. and Kirkland, J.J. (1978) J. Chromatogr. *149*, 465–487.

Yandrasitz, J.R., Berry, G. and Segal, S. (1981) J. Chromatogr. *225*, 319–328.

Yarmchuk, P., Weinberger, R., Hirsch, R.F. and Love, L.J.C. (1982) Anal. Chem. *54*, 2233–2238.

Yau, W.W., Gennard, C.R. and Kirkland, J.J. (1978) J. Chromatogr. *149*, 465–487.

Zakaria, M. and Brown, P.R. (1981) J. Chromatogr. *226*, 267–290.

Zakaria, M. and Brown, P.R. (1983) J. Chromatogr. *255*, 151–161.

Zaspel, B.J. and Saari-Ccallany, A. (1983) Anal. Biochem. *130*, 146–150.

Ziltener, H.J., Chavaillaz, P.-A. and Jorg, A. (1983) Hoppe-Seyler's Z. Physiol. Chem. *364*, 1029–1037.

Subject index